ELEMENTARY PARTICLES

Building Blocks of Matter

ELEMENTARY PARTICLES

Building Blocks of Matter

Harald Fritzsch

University of Munich, Germany

Translated by

Karin Heusch

World Scientific

NEW JERSEY · LONDON · SINGAPORE · BEIJING · SHANGHAI · HONG KONG · TAIPEI · CHENNAI

Published by

World Scientific Publishing Co. Pte. Ltd.

5 Toh Tuck Link, Singapore 596224

USA office: 27 Warren Street, Suite 401-402, Hackensack, NJ 07601

UK office: 57 Shelton Street, Covent Garden, London WC2H 9HE

Library of Congress Cataloging-in-Publication Data
Fritzsch, Harald, 1943–
   [Elementarteilchen. English]
   Elementary particles : building blocks of matter / Harald Fritzsch.
    p. cm.
   ISBN 981-256-141-2 (alk. paper) ISBN 981-256-408-X (pbk, alk. paper)
    1. Particles (Nuclear physics)--Popular works. 2. Quantum
chromodynamics--Popular works. 3. Quantum electrodynamics--Popular
works. I. Title.

QC793.26 .F7513 2005
539.7'2--dc22   2005044181

British Library Cataloguing-in-Publication Data
A catalogue record for this book is available from the British Library.

Printed in Singapore by World Scientific Printers (S) Pte Ltd

Contents

Chapter 1

Introduction

Philosophers in Antiquity were already tackling the question: what does matter surrounding us ultimately consists of? Will we hit a boundary when we take a piece of material, say, a piece of wood or a diamond, and break it down into smaller pieces? If so, what defines that boundary? Are there minuscule objects that can no longer be split up? Or is that boundary simply that a further division, for whatever reason, does not make sense any longer?

Let me give an analogy, starting from the question: what are the original elements of, say, the German library in the city of Leipzig or, for that matter, of any large library? To find out, we step into that library and find thousands of books. We open a book and discover the building blocks of that book – often more than a hundred thousand words. We recall that there are dictionaries that contain information about all of those words – tens of thousands of them. And each of these words is made up of letters. Our alphabet contains twenty-six different letters. All of the words in the books are made up of these twenty-six letters.

These days, books are often written on the computer. The word processor in the computer writes these letters in terms of sequences of the symbols zero and one – they are digitally produced just like the Morse code which allows the letters to be shown as a series of points and lines:

$$a: .-, \quad b: -..., \quad c: -.-., \quad \text{etc.}$$

All words can be reduced to the level of sequences of "bits" – of the elementary building blocks of information, of 0's and 1's. That is where our search for basic elements ends. It makes no sense to try and dissect the symbols 0 and 1 onto an even smaller level. We have come to the end of the process of dissection. The basic elements of a library have thereby been found — 0 and 1. These are also the basic elements of information — we might say, these are the atoms of the science of informatics.

Now let's replace the library by our material universe, the books by what the universe is made up of — objects living or dead that we see around us: trees, rocks, stars, planets, desks, diamonds, ants, dogs....

Every attentive observer of natural phenomena is impressed of the enormous variety of phenomena and of material objects that are observed. Obviously, there is no random set of observed phenomena. Many of these phenomena are repetitive. A speck of salt here looks just like another one thirty miles away, they are as indistinguishable as two eggs. The leaves of a maple tree in Berlin or in Washington D.C. are indistinguishable. On one hand, there is an incalculable variety of phenomena and occurrences; on the other hand, there are also elements that we encounter again and again. This duality of great variety and of constancy prompted the Greek philosophers, thousands of years ago, to set up the hypothesis that the universe consists of elementary building blocks that can no longer be split up the "atoms" (derived from the Greek word "atomos" which means indivisible). Leukippos of Miletus and his student Democritus of Abdera in the 5th century B.C. came up with this hypothesis. Only few such atoms should be enough to produce this enormous variety of things in ever new combinations of atoms. "There is nothing" Democritus said, "except atoms and empty space".

Another viewpoint was presented by Anaxagoras — who was born about 500 B.C. He assumed that there are infinitely many basic materials that, appropriately mixed, can make up the great variety of materials that exist in the world. These materials were assumed to be indestructible and their variety was thought to be due to their motion during the mixing process. Empedocles, some ten years younger than Anaxagoras, felt the materials of our world are basically made up of these four elements: earth, water, air, and fire.

It is interesting that only in the framework of the notion of atoms, the concept of empty space became important. Up to that time it was believed that space was filled with matter, constructed by matter, and there was no thought of empty space. Only in the framework of atomic theory was there an important function for empty space: it became the framework of geometry; the atoms moved around in it.

Atoms move in space but have only geometric qualities. Democritus said: "Just as tragedy and comedy can be written down with the same letters, different phenomena can be realized in our world by the same atoms, provided they assume different positions and different movements. Some given matter may have the appearance of a given color; may appear to us as tasting sweet or bitter – but in reality there are only atoms and empty space".

Later on, Greek philosophy took on the elements of the theory of atoms and developed it. Plato discussed in his dialogue "Timaios" possible connections between the atoms and Pythagoras' model of harmonic numbers. He identified the atoms of the elements: earth, water, air and fire with the regular geometric patterns of dice, octahedrons, icosahedrons and tetrahedrons. As the atoms move, that motion has to be ascribed to the laws of nature: they are not moved by forces like love and hatred, but rather as a manifestation of the laws of nature.

What began about two and one-half thousand years ago in what was then Greece on the Western coast of Turkey, was nothing less than the start of a revolution that is still in progress. For many thousands of years before, people believed that whatever happens in our world is primarily due to supernatural origins. Magic and superstition ruled the world. This changed about two thousand years ago on the Ionic Coast. The time and the location of that emerging turning point were not coincidental. The city states on the Ionic Coast had developed democratic values. New ideas were accepted and were easily spread, not least due to the transition from hieroglyphic writing to scripture based on an alphabet. Religion did not play a dominant role. This permitted the establishment of the idea that our world is recognizable, that the natural processes we observe can be analyzed and understood rationally. The notion of atoms was basic to this development. The red thread that can be followed from the Ionic Coast of Turkey through two-and-one-half millennia of our history to the point where it ends in our present scientifically and technically developed world; and that means, it defines man's knowledge of the building blocks of matter.

Many details of what was taught about atoms in antiquity were revealed to us by happenstance, as it turns out, in the year 1417 when a manuscript of the Roman poet and philosopher Lucretius was found. In well-measured hexameters, this work, called "De rerum natura" (about the nature of matter), Lucretius' described not only ideas originally developed by Leukippos and Democritus; but also developed them further. In the work by Lucretius, ancient atomism reached its highest point. His book was one of the first published after the invention of printing and was circulated throughout the Europe; ever since, it has influenced developing notions of Nature.

With astonishing clarity, Lucretius anticipated the basic elements of modern particle physics. In his manuscript: "De rerum natura" which we might well translate into "The world of atoms", he writes:

"There is no doubt an ultimate end to the body that our senses cannot see; it is not further divisible, minuscule in size, and has never existed all by itself, nor will it be able to do so in the future. It is, after all, part of this body, and resembles it. Since it cannot exist alone, it is rooted in the larger structure, and no power can pull it away. That means there are basic structure elements of the tightest simplicity; they are stuck, tiny as they are, tightly together with their likes – and form a large multiple space together. Not that they form a multiple whole; rather, the forces that hold them together are enriched by their eternal simplicity; nothing can pull them apart, nor can their basic nature be simplified and still hold on to their seed." (De rerum natura, book one).

It is interesting to follow Lucretius' argument for the indivisibility of atoms: *"If there is no such thing as a smallest unit, that would mean that even the tiniest body, or structure, consists of many limitless parts: halving half of it will become part of an unstoppable chain. Where then is the difference between the universe and the tiniest constituent? There is no difference. After all, even if the summation process goes on and on endlessly, the smallest part remains divisible interminably"* (De rerum natura, book one).

Lucretius' writings give us the best and the most detailed account of the atomic teaching in antiquity, which, however, was not able to resist the idealistic teaching of Plato and Aristotle. He brings in the interminable questions of an attempt to understand nature, trying to clear up unending questions about nature; moving away from its mythology without losing a deep respect towards its unshakable laws. Had Lucretius' notion of how to understand nature been accepted two thousand years ago, the course of world history would have looked quite different from what actually happened; certainly it would have avoided religious excesses, and religious wars in Europe and Asia. Unfortunately, reality was different. After the collapse of the Roman Empire, the Western world suffered more than a millennium of intellectual frustration while religious fanaticism and superstition dominated. It was not before the time of the Italian Renaissance that, in large parts of Europe, the intellectual clarity of the Greek way of thought reemerged from its thousand-year atrophy. At that time, scientific thinking made its breakthrough; Copernicus, Johannes Kepler, and Galileo Galilei carried it forward heroically.

In the 17th century, the concepts of philosophers in Antiquity were linked to those of the natural sciences of that time. Scientists of that period noticed that chemical elements such as hydrogen, oxygen, or copper are made up of identical atoms for each species. Isaac Newton (1643–1727), who first wrote down the laws of mechanics and thus became the founding father of theoretical physics, noticed and taught that the stability of materials, such as the cohesiveness of a metal, is based on forces acting between individual atoms.

In the second half of the 19th century, this concept of the atomic basis of material led to important progress in chemistry. The chemists noticed that chemical reactions are not easily understood if we assume that materials in these processes consist of minuscule and not-further-divisible particles: the atoms. A chemical element like hydrogen then consists simply of nothing but one and the same species of atoms. It was possible to determine the size of the atoms by chemical methods; and that turned out to be 10^{-8} cm.

One billion hydrogen atoms might be lined up so as to form a chain 10-cm long. Today, we know one-hundred-and-ten different elements, and that means we know one-hundred-and-ten different kinds of atoms. The Greek philosophers would not be able to subscribe to this idea — they never thought of the possibility that there might be more than one-hundred different kinds of atoms. In the early 20th century, the realization that atoms are not elementary in the sense conceived by the ancient Greek atomists was not due to the chemists; it was the physicists who realized that atoms are not really elementary particles, but rather that they are made up of smaller constituents: of electrons that orbit around atomic nuclei, where the nuclei contain most of the mass of the atoms.

The great breakthrough of atomic physics came in the 1920s and 1930s: the new insight provided by quantum mechanics made it possible for physicists to understand, for the first time, the structure of atomic matter that is governed, both qualitatively and quantitatively, by a few basic principles. Most of the problems that physicists and chemists of the previous centuries had not been able to explain, were now open to elegant explanations.

Next, the physicists applied those very principles to atomic nuclei; they hoped to be able to find a deep understanding for nuclear phenomena as rapidly by following similar lines of thought. That hope turned out to be unrealistic. True enough, it was soon found that atomic nuclei are by no means elementary, but that they consist of smaller particles which we call protons and neutrons. This finding alone did not permit to draw conclusions about the detailed properties of atomic nuclei.

Things got worse: experiments that tried to find the inner structure of atomic nuclei by breaking them up, showed that new particles now emerged — the existence of particles had not been imagined before. A whole range of new-fangled particles was found in a confusing variety, leading some physicists to despair of this confusing array; it made them compare sub-nuclear physics with botany. In the late 1940s, experiments produced many particles which are not actual building blocks of matter; rather, they are needed to hold the building blocks together and to see to the joint operation of the forces of nature.

The breakthrough came in the decades from 1960 to 1980; that is when it became possible to establish some order in the chaos of sub-nuclear phenomena. In the distant future, the 20th century will be remembered as the time in which most of the questions concerning sub-structure of matter were solved. We now know that stable matter consists of quarks – the elementary building blocks of atomic nuclei and of electrons. In the 1970s, all this was incorporated in a clear concept of the microstructure of matter. It was given the somewhat prosaic name the "Standard Model of Particle Physics". This model is the subject of what we want to discuss in the following.

The Standard Model, it turns out, is much more than a theoretical model of elementary particles and their interactions. It lays claim to being a complete theory of all the phenomena that are observed in the realm of elementary particles. Today's experts are able to define it in just a few lines of basic concepts; which constitute a kind of "world formula". Such eminent theoretical physicists such as Albert Einstein and Werner Heisenberg had looked for it, but did not find it. Should we find that this theory – the Standard Model – is representative of the ultimate truth, then we will know that recent progress in physics has been able to define electrons and quarks as truly elementary constituents of nature. It would mean that the atoms of Democritus and Lucretius were truly found. To this day, however, this question has not been answered. The Standard Model contains elements that look unsatisfactory to many physicists. We therefore assume that the model provides no more than a very useful approximation of the true, ultimate theory. Today's experimental physicists are very busy looking for new phenomena, phenomena that we call "beyond the Standard Model".

The electrons and the quarks are not simply the building blocks of matter which could be combined in any way we want. They are subject to forces, such as those of electromagnetism; and these forces in turn are mediated by minuscule particles. That is why, in particle physics, we should not speak of fundamental forces that act between the particles; we should speak of interactions. It turns out, as it were, that these interactions follow well-determined rules in the Standard Model, rules that are dictated by symmetry. A good part of this book will be discussion of these forces. A detailed understanding can be established only in mathematical terms.

The symmetries of nature are tightly connected with the interactions of elementary particles. In ancient times, Plato had already hinted at this connection. Werner Heisenberg, one of the founders of quantum mechanics and one of the most important physicists of the 20th century, put it this way: *"What is an elementary particle?" "Plato did not simply accept elementary particles as a given basis of matter, unchangeable and indivisible. There needs to be an explanation — the question after the existence of elementary particles winds up as a mathematical problem .... The ultimate basis of what we see is not just matter: it is the law of mathematics, its symmetry, its formulation"*.

Does that mean that, ultimately, ideas are more relevant than matter? Or, will the difference between matter and formalism vanish as we determine the ultimate boundaries of particle physics? To this day, this question has not been answered; we cannot even be certain whether the question is truly relevant. At the end of this book we will come back to this point.

Particle physics finds itself challenged by new questions at the beginning of this millennium, at the threshold of new discoveries. Physicists penetrated ever more deeply into the phenomena of basic matter in the 20th century; they opened up a world of new phenomena hitherto unknown, new horizons. The structure of microcosms became visible and turned out to be a complex world all by itself. Fortunately, it was possible to describe it in astonishingly simple mathematical terms. The questions to be answered became even more fundamental: whence does matter come our way? Whither does matter go in the distant future? Particle physics is and remains a great adventure. Every time a new experiment is started, after years of complicated preparations by our experimental colleagues, we all head for a voyage into "terra incognita" (unknown territories).

The distance between the world of particle physicists and the general going-on of our society's daily world is becoming dizzying. After all, why should we see it as our task to find new particles and new phenomena that appear irrelevant for our daily lives? The motivations are on the same level as those that make other scientists explore the universe, the ultimate depth of the ocean, any set of boundaries we have not penetrated yet. Similarly to every basic research activity, particle physics is part of our culture, setting itself the goal of understanding cosmic order. Basic research in physics is mostly concentrated on the particle front. Vast investments made by our society in the past have borne fruit in our understanding as our society became open and enlightened in most parts of the world. Particle physics mediated fascinating insights we gained into the structure of the microcosm. This very fact makes up part of the unquestionable conquests of humanity in the 20th century.

Chapter 2

Electrons and Atomic Nuclei

The history of natural science probably never knew a time where important discoveries were as densely packed as in the last five years of the 19th century; all of these discoveries deal with phenomena that are part of today's particle physics. First, there was the 1895 discovery of a type of radiation which is now generally known, in English as X-rays, whereas the German-speaking world calls it Roentgen radiation after its discoverer Wilhelm Konrad Roentgen (1845–1923). In the following year, Henri Antoine Becquerel (1852 1908) hit on the phenomenon of natural radioactivity, which we will discuss later on. Let us just mention right here that this means the emission of particles by radioactive atoms, with energies a million times above those imparted to atoms in chemical reactions.

Physicists rather than their chemist colleagues noticed, at the beginning of the 20th century that atoms are not really elementary such as the ancient Greeks had believed, but rather that they, in turn, are built up out of smaller building blocks. This notion first appeared when the electron was discovered in 1897 — the first really elementary particle. A number of physicists were involved in this discovery: the Germans Emil Wiechert and Walter Kaufmann, and their British colleague Joseph John Thomson. Initially, the electrons were described as "electrical atoms" in January 1897, by Emil Wiechert in a presentation at Koenigsberg University in East Prussia. The birth date of the electron is frequently quoted in the English-speaking world as April 30, 1897, when Thomson presented his experimental results with cathode rays at the Royal Institution in London.

Thomson was not the only scientist who, at that time, ran experiments on electrical discharge processes between two electrodes, an anode and a cathode, inside glass tubes that had been pumped empty of all air. Such discharges were frequently shown on public fairgrounds to the amazement of onlookers. Those lay people were just as astonished as the physicists at the many complex light phenomena in those tubes. Today's neon tubes are just one successful application. Once we diminish the gas pressure in the glass tube, those light phenomena inside the tube disappear. Then, on the glass surface opposite the cathode, which is connected to the negative voltage, we observe what we call fluorescence. This is a phenomenon due to radiation issued from the cathode, which was therefore originally called cathode rays; but it was not known what kind of radiation it was. It was noticed that it is useful for a number of informative experiments: it heats up materials it finds in its way; it blackens photographic plates. It also influences chemical reactions. And soon enough it was noticed that once those cathode rays hit materials in their path, they leave a negative electrical charge.

Thomson, Kaufmann, Wiechert, and many other physicists performed relevant experiments. Now knowing that these rays carry negative electric charges, Thomson speculated how he might identify them. He directed those rays through electric and magnetic fields and realized how those fields changed the path of his cathode rays. The degree to which their path was changed, he knew, depended on the electric charge and the mass of the "ray". Thomson then noticed that he was observing a constant ratio of charges and masses, irrespective of the medium — air, hydrogen or carbon dioxide. He finally succeeded in measuring the energy, and thereby the mass of the particles: they were lighter by a factor of about 2000 than the hydrogen atoms (to be precise, by a factor of 1837). Thomson then declared he must have hit on a new state of matter — a state that must be part of the makeup of chemical elements. And it turned out that his hypothesis was correct: these particles which were later on called electrons are important ingredients of atoms. Their name is due to the Greek word "elektron", which described a silver-gold mix, or amber).

Walter Kaufmann in Berlin was another person to do experiments with cathode rays: he determined the ratio of the particles more precisely than Thomson. But, unfortunately, he was strongly influenced by the Vienna physicist and philosopher Ernst Mach. Mach refused to introduce hypothetical objects such as atoms or even smaller particles if they were not directly observable. That is why Kaufmann avoided talking about the discovery of new particles. Had he done so, we would have to recognize him, together with Thomson, as the discoverer of the electron. Let this example simply show that, in physics as in other natural sciences, it is not only important to discover a new phenomenon, but also to come to a correct interpretation. For the latter it is very important to advance the fitting theoretical framework.

So, what Thomson found was nothing less than the lightest electrically charged elementary particle. Electrons are stable; they have a negative charge and a mass of $9.1093897 \times 10^{-31}$ kg. Calling its charge "negative" is simply a convention. It was introduced by the American scientist and politician Benjamin Franklin in the 18th century. In retrospect, we might say, this was not the most fitting choice: after all, electric currents are due to electrons that propagate inside metal. So, it would have made sense to call the charge of the electron, the most important charge carrier, positive.

The charge and the mass of the electron are important constants of Nature — they are exactly the same everywhere in the universe. We usually denote the charge amount of the electron by the symbol, e, since all electrons carry precisely the same amount of charge, we say: the electric charge of the electron is quantized.

As far as the electron mass, and the masses of all other particles are concerned, it would make no sense to measure them in the unit kg or g, which makes no practical sense in microphysics. We tend to measure it in terms of energy units, specifically in electron Volts. Recall that Albert Einstein in the year 1906, discovered the equivalence of energy and mass, which he expressed in the famous relation $E = mc^2$; this suggests expressing the mass of all particles

in units of electron Volts. One electron Volt or, for short 1 eV, is
the energy given to an electron as it is accelerated by an electric field
measuring 1 Volt strength. If we use a voltage source of 1.5 Volt to
charge up two parallel metal plates, an electron that is very close to
the negatively charged plate will be accelerated in the direction of
the positively charged one; and as it arrives there, it will have kinetic
energy of 1.5 eV.

In many specific contexts of particle physics, the unit eV is too
small. Depending on the specific application, we then use units such
as kilo-electron-volts (where 1 keV = 1000 eV), Mega-electron-volts
(1 MeV = 1000 keV), Giga-electron-volt (1 GeV = 1000 MeV), and
Tera-electron-Volt (1 TeV = 1000 GeV). To give an example, the electrons that generate images on computer or television screens have energies of a few times 10 keV. The LEP accelerator (the so-called Large
Electron Positron collider) at the European research center CERN
was able to accelerate electrons up to energies of about 110 GeV
until its demise in the year 2000. And Einstein's mass–energy relation lets us express the mass of an electron in energy units also:
$m(e) = 0.511$ MeV.

When it comes to practical applications, the electron is the most
important elementary particle. Ever since its first discovery, it has
presented important challenges to physicists. It has mass, we know,
and that suggested it must also have a finite size, to be expressed by a
definite radius. But experiments that are being performed with ever
greater accuracy tell us that the radius is zero. That would make it
as elementary an object as we can imagine: a mathematical point
that does possess both a charge and a mass. Should the electron
have any inner structure, that structure would have to be smaller
than 10^{-17} cm — a billion times smaller than the size of an atom.
Is it possible that there is a pointlike object that has both mass and
charge? Mathematically speaking, a point is the result of a limiting
definition: it can be imagined as a very small sphere of radius zero.
We will have to imagine a tiny sphere with the mass and the charge
of an electron, the radius of which we diminish until it reaches zero.
This makes the electron a mathematical construction, sort of like
Lewis Carroll's Cheshire Cat which, you may recall, vanishes slowly

while leaving behind nothing but its smile: a catless smile. Let us wait and see whether future physicists who do experiments with electrons will lose their smile in the process.

Having discovered the electron, Thomson speculated that essentially all of atomic matter consists of its particles that are embedded in a somewhat diffuse, electrically positive charged medium; but this time he did not hit the nail on the head. Quite early in the 20th century, it was noticed that the mass of the atoms is mostly not tied to the electrons, but is due to a "massive nucleus". It was Ernest Rutherford who, with his group at the University of Manchester in Great Britain, performed the decisive experiment in the year 1909. They used alpha rays emitted by a radioactive substance — even knowing that these doubly charged particles are simply nuclei of the element helium. These alpha rays traversed a gold foil in an experimental setup that was embedded in zinc sulfate casing. When an alpha particle hits a zinc sulfate molecule, it causes a small lightning to be emitted, visible to the human eye in the dark laboratory. In this fashion, it became obvious when an alpha particle, having penetrated the foil, had a close interaction with one of the atoms which scattered it such as to change its direction. Most of these scatterings happened at very small angles, but once in a while they led to a change of direction by 10 degrees or more.

Rutherford and his collaborators performed a very systematic investigation, trying to find out whether, once in a while, an alpha particle might penetrate the gold foil only to turn around and fly back — sort of like a tennis ball that bounces back from the wall. Nobody expected this to happen but, amazingly, it did. Roughly one in 8000 alpha particles did just that to the astonishment of the physicists in this group. Rutherford would recall later on: "This was the most incredible result of my related work that I experienced. It was as though we took a 15-inch grenade and shot it against a paper foil, only to see the grenade come back at the person who launched it".

It took almost two years for Rutherford to make sense out of this phenomenon. There was only one way to understand this odd backscattering of the particles. Essentially all of the mass and the

positive electric charge of the gold atoms in the foil had to be concentrated in a very small volume way inside the atom, making up its
nucleus. But the electron which mass makes up only a tiny fraction
of the atoms, moves rapidly through the atomic volume. Further
experiments permitted an estimate of the size of the nucleus. It
turned out to be about 10,000 times smaller than the atomic radius,
or some 10^{-12} to 10^{-13} cm. And that indicated that the volume of
the nucleus was laughably small compared with that of the atom.
The atoms were mostly just empty space. When we touch the surface of a diamond, we gain the impression that it is a very dense bit
of matter. And so it actually is — but its very hard appearance is
due to a combination of electrical forces acting between the nuclei
and the electrons. A diamond ultimately consists of mostly empty
space; alpha particles can penetrate it as easily as they pass through
a gold foil. Let us recall the ancient saying of Democritus: "There
is nothing in existence except atoms and empty space". He would
be extremely astonished if, in a modern reincarnation, he found out
that the atoms also consist of mostly empty space.

Rutherford himself often pointed out that good luck is one of the
necessary ingredients for experimental physicists when they make an
important discovery. Ultimately, it turned out much later that this
was in fact the case in his experiment on alpha rays. As luck would
have it, Rutherford was using a radioactive source that emitted alpha
particles of 5 MeV — an ideal energy for the discovery of atomic
nuclei. Had the energy been higher, the alpha particles would have
engaged in complicated reactions with atomic nuclei. That would
have made it impossible to find results which were easy to interpret.
If, on the other hand, their energy had been lower than 5 MeV, the
backscattering would not have been observed in the first place.

Rutherford was able to demonstrate that the scattering of alpha
particles followed a simple rule: the particles were positively charged,
as were the atomic nuclei. The scattering therefore was due solely
to the electrostatic repulsion of these two systems. Inside the atoms,
it was shown that the well-known law applied: equal charges repel
each other, opposite charges attract each other. Had Rutherford performed his experiments with electron beams, he would have observed

similar scattering phenomena, with the difference that the electrons and the nuclei attracted each other. In those days, however, there was no way to generate an electron beam with energies of about 5 MeV.

Rutherford's model of atoms started a very rapid development, leading up to the model of atomic structure that we consider to be correct today: atoms consist of a positively charged nucleus surrounded by a shell of negatively charged electrons. The positive charge of the nucleus, of course, is due to protons.

The simplest atom we know of, the hydrogen atom, has just one electron providing this shell for one proton. Clearly, nucleus and electron attract each other, and it is only the steady almost circular motion of the electron that keeps it from being pulled into the proton. We might think of the hydrogen atom as of a microscopic planetary system: its center is — taking the role of our Sun — the nucleus consisting of just one proton; it is steadily circumnavigated by the electron, the stand-in for our Earth. The electron follows a circular path around the Sun — and electrostatic attraction is being cancelled by the centrifugal force which opposes it.

The hydrogen nucleus is just one proton, which permits us to study the properties of this particle in detail. Its mass was determined to amount to 938.272 MeV. This makes it almost precisely 1836 times as heavy as the electron. This atom therefore concentrates more than 99.9% of its mass in the nucleus, to this day, we do not know why the mass ratio of electron to proton is so minuscule. After all, the electric charge of a proton is precisely the same as that of the electron, once we change its sign — it is simply e. The electric charge of the atom is therefore exactly zero. This is another phenomenon we do not fully understand; we might assume that the absolute value of the electron and proton charges is slightly different in addition to their opposite signs. That, of course, would make the charge of the hydrogen atom different from zero.

The fact that hydrogen atoms are devoid of electric charge is extremely well established. In the universe, we observe large systems of matter, such as gaseous clouds, that consist mainly of hydrogen; that fact permits us to write down very precise limits for any assumed non-zero charge of the hydrogen atom. It has to be smaller than 10^{-21} times that of the proton. This limit can be set with such precision because any non-zero charge of the hydrogen atom would lead to electrostatic repulsion in hydrogen clouds. But such a repulsion has not been observed. We therefore have this remarkable result of equal but opposite charges of proton and electron, notwithstanding the fact that the physical properties of these particles are vastly different — as their extreme mass ratio makes obvious. It does tell us that there must be something the electron and proton have in common — something that mandates the equality of the absolute value of their charges. We will find that there may well be a common property at the basis of this charge phenomenon. But that is well beyond the realm of atomic physics.

All atoms, with the singular exception of hydrogen, have more complicated nuclei consisting of several particles. We might start with a special kind of hydrogen — often called "heavy hydrogen" — with an atomic nucleus of roughly twice the mass of normal hydrogen, but with the same electric charge. Its nucleus contains a second particle, a neutron, the existence of which was discovered in 1932 by the British physicist James Chadwick. Its mass was determined to be 939.565 MeV, which is about 1% heavier than that of the proton.

The fact that the masses of these two nucleons are almost the same is not coincidence. We will see that this is a consequence of special dynamical properties of theirs. Up-till now, we do not really understand why the neutron has a slightly heavier mass than the proton — it is one fact of microphysics that we cannot really explain. Initially, we might expect the opposite because the proton has an electric charge, and is therefore surrounded by an electric field. Its energy might be expected, according to Einstein's mass/energy relation, to contribute to the mass of the proton.

The neutrons and the protons, often called nucleons, are the building blocks of all atomic nuclei. The relative numbers of neutrons and protons depend on the individual elements of interest. Take, for instance, the carbon nucleus: it usually contains six protons and six neutrons. Heavier nuclei have more neutrons than protons; e.g., the uranium nucleus has 92 protons and, usually, 146 neutrons. Inside the nucleus, the electrostatic repulsion between the protons would cause this compound to explode were it not for a new force of nature, the so-called strong interaction. This force is attractive between the nucleons and sees to it that protons and neutrons co-exist in the smallest space. The diameter of atomic nuclei is of the order of 10^{-13} cm. That makes nuclei about 100,000 times smaller than the atom as a whole. Were we able to inflate the atomic nucleus to the size of an apple, the atom would have to attain the respectable diameter of about 10 km.

What is the basic need for the existence of this so-called strong force? Until the 1970s, this was an unsolved problem. Clearly, it has to be connected with the structure of the nucleons. Unlike the electrons that surround them, the constituent particles of the nuclei are not devoid of a structure of their own. They do have a sub-structure, as the atoms do in a way that we described above. Before we attempt to tackle this question for the nuclei, we have to deal with another important property of microphysics that is of decisive importance for the sub-structure of matter: the quantization of microphysical phenomena.

Chapter 3

Quantum Properties of Atoms and Particles

It is obvious that all atoms share the same basic structure, one hydrogen atom resembles every other hydrogen atom. This property of atoms illustrates the duality of riches and of uniformity displayed by Nature, as we said earlier on. But classical physics does not explain this feature, since its laws have no basic scale of things. Classically, there is no reason why a given electron should be at a distance of one millionth of a centimeter from the nucleus it is bound to, or rather at a thousand times that distance or a million times farther still. So what could be the reason why the electron, bound in a hydrogen atom, goes at a speed such that its distance from the nucleus is always the same, i.e., about 10^{-8} cm?

In the same viewpoint, classical physics provides no explanation why atoms are stable particles: the electron has an electric charge that moves around the nucleus; so we might expect it to behave like an electromagnetic transmitter which radiates off energy in the form of electromagnetic waves. Now, the diminishing energy of the electrons would then cause the electron to drop into the nucleus — but no such thing happens. There has to be some mechanism that makes the electron move around the nucleus in a stable orbit.

It is quantum theory that provides us this mechanism, giving us the theoretical framework which permits a description of "microphysics" (physics at very small distances). In the world of atoms and subatomic particles, the well-known laws of normal mechanics, for which we have an intuitive understanding, no longer apply. While the German physicist Max Planck developed the first ideas of quantum mechanics in the year 1900, it took more than two decades until the implications of these ideas became clear thanks to the work of Arnold Sommerfeld, Niels Bohr, Werner Heisenberg, and Wolfgang Pauli. But to this day it has remained a mystery why quantum theory is able to describe microphysics so well. My ex-colleague at Caltech, Richard Feynman, himself a leading quantum theorist, often said: "Nobody understands quantum theory." Niels Bohr, one of the first luminaries, put it differently: "Nobody", he said, "understands quantum theory without turning dizzy".

And indeed, quantum theory breaks the concepts of physical phenomena that every one of us has developed during his or her life: concepts that are deeply rooted in our intuitive understanding of physical reality become irrelevant in this framework. Microphysics permits processes that our understanding of classical physics excludes. It appears that we are not able to conceptualize the dynamics of microphysics in terms of the concepts that have evolved over many centuries. But still, quantum physics permits us to calculate observed processes with spectacular precision, even though a deep understanding has not really been achieved.

Quantum theory is defined around a tiny constant of nature, the quantum of action called h, unknown in classical physics. It has experimentally been determined to have the value 6.6×10^{-36}. Watt seconds: clearly a tiny value when measured in the usual units of Watts and seconds. That is the very reason why the astonishing phenomena of quantum physics do not show up in our daily lives. The name – quantum of action – is due to the fact that this constant gives a measure of physical action, where "action" is defined as a product of energy and time. This makes intuitive sense: the action of some happening can be seen by having a given energy available for a given amount of time. If that time is very short, this "action" will be very small even if the energy in the process is sizable.

One of the important implications of quantum physics says that the physical quantities that need to be known for the description of the motion of an electron around the nucleus, such as its location and its velocity, can never be measured precisely; rather, there always remains an uncertainty which is defined by the "uncertainty relations" first noticed by Werner Heisenberg. Consequently, we must give up on trying to come up with a really exact description of the dynamics inside an atom. The only quantity we can determine is the likelihood that a given process with given parameters is happening: it is impossible to find out where an electron is at a given time, and how fast it is moving. If we concentrate our interest on some exact location of an electron, we have to give up on finding its velocity and vice versa. How indeterminate the product of location and velocity is will be given by the quantum of action. Note that such relations of lacking precision are also known in the macroscopic world — say, in the case of a moving car. But the uncertainty relations imposed by quantum theory on location and velocity of the car are so minuscule that we can safely neglect them, and well below what can be measured. That is the reason for our lack of an intuitive understanding we might have for the quantum aspects of the world around us.

In atomic physics, things are very different. It is just this "uncertainty" that determines the size of, say, a hydrogen atom. The uncertainty in the location of the electron in the atom is the same quantity that we call the diameter of the full atom which includes the electron orbits. It amounts to some 10^{-8} cm. Now assume that we look at a hydrogen atom with much smaller electron orbits — say, a hundred times smaller than the one we know from stable atoms. This would mean the electron is much more precisely localized than in the natural world: the uncertainty relation defines a much larger range of velocities for the electron than in the normal atom so that, on the average, the electron will move much faster than in the "normal" atom. But higher velocities mean higher energy so that the smaller atom is more energetic by comparison. This, however, runs counter to an important principle of Nature — namely the fact that every system, including an atom, always tries to be in a state of lowest energy. Consequently, the "smaller" atom will be unstable. It will soon radiate off energy and assume a larger size until it looks like a normal atom.

In the same way, we can look at an artificial atom that has a hundred times the size of a natural atom. To create it, we would have to pull the electron away from the nucleus, and to do this, we need energy. Consequently, the energy of this new atom exceeds that of the "normal atom". This larger atom will also, soon enough, make a transition to the "ground state". This is the state of minimal energy. There is no way to coerce the electron into giving up even more energy. It is then the uncertainty relation between location and velocity that determines the size of the atom.

To be precise, it is not the electron's velocity that shows up in this uncertainty relation — rather, it is its "momentum", the product of velocity and mass. That means that the size of the electron "shell" around the nucleus depends on the electron mass. If that mass were smaller by a factor of one hundred, this shell would be larger by the same factor — that would give it a diameter of one millionth of a centimeter. If, on the other hand, it were bigger by a factor of ten thousand, the shell would have a diameter larger than that of the nucleus by only a factor of ten. This implies that the size of the atom as enforced by quantum theory gives a major element of stability to Nature. And thus quantum theory gives us an explanation for Nature's tendency to come up with universal patterns in atomic physics, in chemistry, and in biology, notwithstanding all changes she may have to accommodate.

Quantum mechanical uncertainty makes it impossible to "follow" the electron's motion around the nucleus. Not only that, it makes no real sense to try and define a given orbit: we can only determine the likelihood that the electron is in some given space close to the nucleus. This likelihood distribution does not at all look like an orbit — rather, it surrounds the nucleus with spherical symmetry — and its maximum is in the very center of the atom, where we know the proton is. It is defined by the wave function of the electron — and that wave function can be precisely defined by quantum theory. It describes the state of the atom fully. It sounds odd, and so it is: quantum theory does not give us information on facts. It tells us about the likelihood — but that information turns out to be precise.

There is another feature of quantum theory that manifests itself in particle physics: the existence of excited states. Say, we add energy to a hydrogen atom by irradiating it with electromagnetic waves: the electron may then jump to another state of higher energy. Such excited states have specific energies. Physicists speak of a discrete energy spectrum, the lowest state of which is called the ground state of the system. The irradiation makes the electron jump up to a higher energy state: but it will revert to the ground state after a short time, emitting electromagnetic radiation in the process. This whole process can occur only if the atom absorbs precisely the energy needed for the excitation we mentioned.

It turns out that knowledge of the excitation energy is not sufficient to fully describe the wave function of the excited atom. We need to know more, such as: what is the angular momentum of the excited atom? In its ground state, we can point it in any direction without changing its energy, because its wave function is spherically symmetric and has no angular momentum. But the excited states do have that possibility. Their possible angular momenta are quantized just like their energies — but in easier-to-determine amounts: angular momenta must be multiples of the smallest non-zero amount which is simply designated by the symbol $\hbar = h/2\pi$ etc. For short, we often speak only of quantized angular momenta $0, 1, 2, \ldots$, leaving the $\hbar$ out.

The ground state of the hydrogen atom has zero angular momentum because the wave function of its electron is spherically symmetric. No direction in space has special significance. This fact alone demonstrates that there are problems with attempts at trying to understand the hydrogen atom with our intuitive notions based on classical mechanics: in that kind of notion, the electron would just orbit around the protons. But every such orbital motion would imply that the electron has some given angular momentum. Angular momentum zero would mean that the electron is at rest next to the proton; that, however, is not permitted by the uncertainty relation, which says that relegating the electron to a tiny space means it must have a large velocity, a high energy. That, again, would mean the atom cannot be in its lowest energy state, the "ground state"!

In the 1920s, physicists noticed that electrons are more complicated than they had thought before. Up until then, the "police file"
for the electron had been comparatively simple: a point-like object
of a well-known mass and electric charge. Once it moved, it acquired
a measurable velocity and, thereby, momentum and motion energy.
While at rest, its only energy was given by its mass according to
Einstein's equivalence of mass and energy, $E = mc^2$. But now it
turned out that the electron had an added "property": an angular
momentum.

Now there is a conflict with classical physics: let's imagine the
electron is a little sphere like, say, a tiny tennis ball; now this may
easily have angular momentum, whether it has linear motion or not.
It may revolve around any axis, even if its center-of-gravity remains
fixed. We then say, it has its proper angular momentum. The axis
around which it turns defines the "direction" of this angular momentum — which has its direction just like velocity or momentum;
mathematically, we speak of a "vector". Suppose we make the radius
of this spherical object smaller and smaller but keep the angular momentum constant (it is what we call a "conserved quantity"), then
the sphere has to revolve faster and faster — just like a figure skater
who revolves in a "pirouette" and pulls in her arms as best she can.
Ultimately, we can imagine that the radius of our sphere is zero: the
sphere is now a point — but the angular momentum remains the
same. That means the angular velocity has to become infinitively
large.

Based on this notion of a limiting effect, we can imagine that the
angular momentum of a "point particle" will have a certain value.
Put that together with our knowledge that quantum theory permits
only certain discrete values of angular momentum to atoms. It is
the same for the internal angular momentum of individual particles,
for which a separate name was chosen: spin. Clearly, spin can have
only certain values — and for the electron that spin was found to be
non-zero: it has one-half the value of the quantity h we mentioned
above. In other words: the electron has spin $\hbar/2$, or $h/4\pi$.

Now we might ask whether it makes sense to interpret this value
of electron spin in terms of an angular momentum that is due to the

internal structure of this particle. That would imply some rotation of matter inside the electron. Attempts to measure such an effect have been totally futile. Contrary to angular momentum that is due to orbital motion, electron spin is purely a quantum phenomenon, devoid of a classical physics parallel. Just as the electron's mass or charge, its spin is an internal degree of freedom; a spin-less electron would make no sense. Recall that we can reduce the angular momentum of a classical object — say, of a rotating sphere — to zero, just by slowing down its rotational motion to zero; but that cannot apply to a particle's spin. The electron's spin cannot be switched on or off — it is a permanent feature of that particle.

In particle physics, it has become common to take the quantity $\hbar$ for granted when mentioning the spin of a particle. We simply say spin-1/2, or, the electron is a spin-1/2 particle. Although spin angular momentum is of different origin from orbital angular momentum, both share the property of being vectors, quantities that have a direction. An electron at rest may have its spin point to an arbitrary direction. But quantum theory tells us that we need to consider only two directions for its spin — any given direction and its opposite. We would then speak of spin +1/2 and spin −1/2. We speak of two different spin states. Should the particle's spin point in a different direction, we can easily construe its spin state as a linear combination of those two previously defined states (Fig. 3.1).

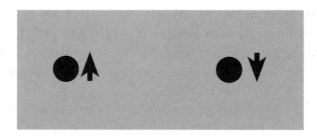

Fig. 3.1. The two different spin states +1/2 on the left and −1/2 on the right.

An electron orbiting an atomic nucleus can be characterized by two facts pertaining to its angular momentum properties: what is its angular momentum? and what is its spin? The first one can only be integers — say, 0, 1, 2,..., the latter is simply $+1/2$ or $-1/2$. Quantum theory does not permit other, more arbitrary, values. The two numeric values we find are called the quantum numbers of the electron.

When detailed studies of the electron "clouds" of atoms became possible in the first quarter of the twentieth century, another peculiarity of quantum physics was discovered, beyond what classical physics could explain: In atomic physics, we might imagine that two electrons orbiting the same nucleus share the same set of quantum numbers — say, orbital angular momentum zero and spin $+1/2$. The laws of quantum physics would permit this coincidence, but it turns out that Nature does not. Wolfgang Pauli introduced another law of quantum physics which says that two electrons in the same atom may not share the same set of quantum numbers. Physicists call this the "Pauli principle". It was found later on that the Pauli principle is due to the electron spin having half-integer values. Were the values integer — say, 0, 1,... — a possibility that is not "forbidden" but is not realized in Nature — things would look different. In that case, two electrons could share the same set of quantum numbers.

There are two "nucleons", protons and neutrons; both have spin-1/2. They also have to abide by the Pauli principle: two protons or two neutrons in one and the same nucleus must not share the same quantum numbers. This fact explains many properties of nuclear structure. But there are other particles that carry integer spin numbers like 0, 1, and so on. We will discuss that later — those particles are unstable; they are produced in collisions and decay shortly thereafter. To this day, we have not found a single massive stable particle with integer spin; there is just spin-zero "particle" of light, the photon — but it has no mass.

Light also has its quantum properties, which is why we speak of "light particles". Isaac Newton, the creator of what we call classical mechanics, told us in the 17th century that light can be described in terms of a current of "light particles". This statement did not find a good resonance, because it turned out that light phenomena could well be described in terms of wave phenomena. This does not necessarily contradict the notion of light particles; after all, everybody has seen waves on the surface of water ponds when we throw a stone into them. Those waves clearly consist of water molecules, of minuscule particles. It makes sense to think of a wave of light in the same way: it is ultimately made up of many minuscule particles, although these particles have no mass: the photons. When we feel the heat of bright sunshine on our skins, this heat is due to the impact of the photons which transfer their energy to our skin.

So, what is the profile of the photon? The photon is tagged simply by its energy, its mass, and its spin — just like the electron. Its electric charge is, of course, zero. But even beyond that, the photon is a very special particle: it has no mass. Einstein's theory of relativity — which we cannot quickly explain in this context — tells us that the absence of mass implies that the photon can never be at rest. It always moves, just like most managers in our society. But unlike all of those managers, it always moves at one ultimate velocity, the velocity of light.

Photons have all kinds of different energies irrespective of their shared mass and velocity; their energy depends on their frequency. Quantum theory tells us that it is simply the product of this frequency and Planck's constant h, the "quantum of action": $E = hr$. This implies that photons of lower frequency — like red light — have less energy than photons of higher frequency — say, blue light. Both of these have less energy than photons of X-rays. The latter carry energies of a few eV (electron Volts) each.

Today, particle accelerators help us to produce photons with very high energies — say, beyond 100 GeV. Radiation made up of photons that are way more energetic than visible light are called "gamma rays"; their individual photons are "gamma quanta". Many unstable atomic nuclei emit gamma quanta as they decay.

A high-energy photon (a gamma quantum) that penetrates into an atom may well interact with the strong electromagnetic field close to the nucleus in a very interesting way, as we show in Fig. 3.2.

In this reaction, two particles are being created from the available energy: an electron and a positron — which has the same mass, the same spin, but opposite charge. This is the antiparticle of the electron. It was discovered in 1932 in an experiment that looked

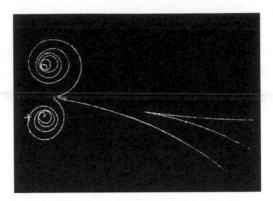

Fig. 3.2.   A nuclear reaction produces a photon which interacts with the electric field of another atom, producing an electron and its antiparticle, a positron. Both of these particles move downward on trajectories that are curved because they traverse a strong magnetic field. The electron trajectory is curved to the left, that of the positron to the right.

into cosmic radiation. Figure 3.2 shows how a nuclear reaction also generates an electron and a positron. It does so by using the energy of the "gamma quantum of energy", the photon, the energy of which is larger than what it takes to create the two masses — larger than about 1 MeV. This does not imply that Nature provides us with an antiparticle for every electron. Rather, it means that an antiparticle can be created in a particle collision. As we explained above, a particle and its antiparticle can be created from energy that must be available in sufficient quantity — a particle pair; and this process is called pair creation.

Just like most processes in physics, the pair creation process can also run in the opposite direction; we call that pair annihilation. When an electron collides with a positron, this collision annihilates both particles, leaving us with two photons. And these photons share the energy of the electron and positron masses plus their motion energy. This process illustrates excellently that matter itself is not indestructible. Matter can be created and annihilated.

Experiments tell us that there is an antiparticle for every particle: protons and neutrons have antiprotons and antineutrons. Antiprotons have a negative electric charge and antineutrons are neutral. Nature around us does not contain these antiparticles, but it is straightforward to create them in particle collisions. There are also particles that are identical to their antiparticles — particles that have no charges of any kind. The photon belongs to this category.

It is by no means a coincidence that particles are created in pairs during collisions. This is due to a symmetry principle which is based on Einstein's theory of relativity and on quantum physics. Particles and antiparticles are completely symmetric. This implies that particles and antiparticles share the same mass. Particle–antiparticle symmetry is the first example of a symmetry that is not due to the inner structure of space and time, but rather to the inner structure of microphysics. That is why this kind of symmetry is also called an "internal symmetry". As we go along in our considerations, we will encounter a number of such internal symmetries.

Chapter 4

The Knives of Democritus

Rutherford investigated the structure of atoms using alpha particles. These particles are easy to come by — there are radioactive nuclei that emit them. But if we want to investigate the inner structure of atomic nuclei and their constituents, alpha particles are completely useless. Why? Because they are nuclei themselves, helium nuclei, and their energy of about 5 MeV would be way too low. It may sound paradoxical, but there is this consequence of quantum theory: the smaller the structure we wish to investigate, the larger the energy and momentum of the particle beams we need for this purpose. Also, this implies the need for larger experimental setups. The uncertainty relation says a high momentum means a small uncertainty, and vice versa.

It is easy to perform a little calculation that shows we can use alpha particles of an energy of a few MeV, such as in the Rutherford scattering experiment, to investigate structures of no more than about 10^{-12} cm. That is about as precise as we can make anything out. If we want to resolve smaller structures, we need higher energy particles. We might say that the knife of Democritus that we need to resolve structures way inside atomic matter, becomes sharper the higher the energy of our beams. For that reason, particle physicists are constrained to build devices that are able to accelerate stable particles such as electrons and protons to very high energies – different kinds of particle accelerators.

Particle physics is therefore frequently called high energy physics, an indication that particle properties cannot fully be investigated except at high energies. Modern particle accelerators are complicated, and very costly devices, notwithstanding the fact that acceleration technologies are straightforward and easy to understand. Most scientists can, if only conceptually, build one of these devices on their own. Take a glass tube just a couple of inches long, two metal plates, and a 12 Volt automobile battery. Fasten the metal plates to the ends of the tube and connect them to the two poles of the battery; now pump the air out of the glass tube. And lo and behold — an electron inside the tube, if it happens to be near the negative metal plate, will be repelled and move, steadily accelerated by electrostatic force, in the direction of the positively charged plate. As it hits that plate, its energy will be 12 electron Volts or, for short, 12 eV. In this simple setup, we can generate a considerable electron current if only we manage to free many electrons from the negative charged metal plate. This can be done by irradiating it with powerful electromagnetic radiation, or by heating up a metal wire in its vicinity so that many electrons can exit from its surface. This is not dissimilar from the tube of a television set, with the important difference that the electrons hitting the image plane carry much more energy — about 20,000 eV.

We might arrange this kind of system such as to change it into a particle accelerator: we replace the metal plate with a thin, flat wire mesh. This attracts the electrons just like the plate did, with the difference that most electrons simply traverse the mesh to make up a beam of electrons of energy 12 eV. This energy is fairly low, but the velocity v of the electrons is still considerable. We can easily calculate it. It is way below the mass of the electrons once we express it in eV, which we know to be 511,000 eV. Now, with motion energy of the electrons being 12 eV, this equals $1/2$ mv^2. We also know that after Einstein's mass-energy relation $E = mc^2$, the ratio 12 eV/511,000 eV $= 1/2$ v^2/c^2. This tells us that the velocity of the electrons is just below 2060 km/s, if we assume that it starts from zero when the electrons leave the negatively charged plate. If their energy were 20,000 eV, as in a television set, we could easily determine the electron velocity to be about 84,000 km/s — which is more than a quarter of the velocity of light.

Every accelerator operates according to the principle we described above. No matter how complicated its configuration in detail, the particles are always accelerated by means of electric fields. This, of course, will function only for particles that are electrically charged. Neutrons cannot be accelerated in this fashion. Also, the particles have to be stable or, at least, possess a relatively long lifetime; particle acceleration, after all, takes its time. This poses no problems in the case of electrons and protons; but most of the other particles that physicists may feed into accelerators, will have to be accelerated within their lifetime.

Atoms, which consist of electrons, protons and neutrons, are a prolific source of electrons and protons for acceleration. Positrons and antiprotons, which we can also accelerate, are not as easily obtained: it takes particle collisions to make them available. It is also possible to accelerate the nuclei of heavy hydrogen, which we call deuterons; or helium nuclei, the alpha particles we discussed above. It is even possible to accelerate the nuclei of heavy atoms such as lead or uranium, to high energies.

The 12 eV accelerator we described above raises the velocity of the electrons to no more than 2,000 km/s. If we add three additional accelerating elements of the same kind, we wind up with the fourfold energy, 48 eV — which means electrons of twice the velocity — well over 4,000 km/s. By adding many similar accelerating elements, we can keep raising the energy, but not the velocity. Once we reach velocities of about 100,000 km/s, the laws of classical mechanics have to be replaced by those of Einstein's relativistic dynamics. The closer we get to the velocity of light, the harder it becomes to raise the velocity of particles. They reach well over 99% of the velocity of light relatively easily, but further acceleration is barely possible, even if we raise the energy. Once we have reached velocities of 0.99 c, the energy will have to be raised by a factor of 3 to reach a velocity of 0.999 c.

Fortunately, it is the energy rather than the velocity that determines the sharp cutting edge of the knife of Democritus. Many years ago, the US Congress was involved in a tough discussion about the funding of particle physicists, and it took a long time to coax approval from the majority. It turns out that they had a hard time understanding why it was important to build an accelerator that would bring the speed of protons not just to 99% of the velocity of light, but to 99.95%, which was going to make it much more expensive. It sounded sensible when some congressmen argued for just 99%, not even a percent less than the value the physicists wanted, but so much easier to obtain.

A machine that accelerated protons to 99% of c, the velocity of light, making their energy 7 GeV, was built in 1955 in Berkeley, CA. It was called the Bevatron; it permitted the discovery of the antiproton, the antiparticle of the proton. At the opposite side of the continent, the accelerator AGS in Brookhaven on Long Island, NY started operations in 1960; it reached proton velocities of 99.95% of c, at an energy of 30 GeV. The so-called Large Hadron Collider, LHC, which is almost ready to go into operation at the European Center of Particle Physics, CERN, in Geneva, Switzerland, will reach a total energy of 14 TeV (14,000 GeV), corresponding to 15,000 times the mass of the proton when expressed in energy units. The proton velocity in this latter case is 0.999999998 c. This demonstrates that the velocity of the particles in high energy physics can no longer be seen as a useful quantity — the relevant quantities are the energy and the momentum of the projectiles.

We should also mention that the statement of Newtonian mechanics that momentum equals mass times velocity is no longer valid in high energy physics. According to Einstein's dynamics, energy and momentum are proportional to each other, tied together by a constant which is the velocity — the velocity of light at high energies. Newtonian mechanics is applicable only at velocities that are considerably lower than c; here, the energy is mostly defined by the mass, i.e., by the relation $E = mc^2$. A proton of energy 100 GeV has a momentum of 100 GeV/c: The momentum of a high energy particle can be seen as directed energy.

In the 1960s, a new accelerator was taken into operation on the site of Stanford University in Palo Alto, CA: the Stanford Linear

Accelerator Center (SLAC), took great pride in producing a new icon of accelerator technology. Here, electrons are being pushed to higher and higher energies by electromagnetic fields along a two-mile linear acceleration line. Originally taken into operation at 20 GeV for energy, this was raised to 50 GeV later on, giving the SLAC physicists a gigantic microscope for an elucidation of the structure of atomic nuclei. We shall discuss the structures that were discovered with this new instrument.

It had been noticed around 1930 that one single element of acceleration can propel particles to high energies. The "trick" is simply that one and the same particle can be accelerated again and again by the identical accelerator element. This, of course, presupposes that the particles revert on a direction, changing, say, on a circular track to their initial location, and are then re-accelerated by the same electromagnetic field. An electrically charged particle in a constant magnetic field will run along a circular orbit. Exploiting the laws of electrodynamics, we can then make the particles pick up additional energy with every full unit of circular motion. If we left the magnetic field constant, the particles would leave their circular orbit after having been accelerated, because the centrifugal force changes their orbit. As their energy increases, the magnetic field has to be increased accordingly. In principle, the particle's energies could be increased arbitrarily, were it not for the need for stronger and stronger magnetic fields. That, however, has its limitations due to several factors, such as the materials used and the particular technology employed.

A circular accelerator, in which particles move inside an equally circular vacuum pipe, are defined by two parameters: the radius of the ring and the maximal magnetic field it can furnish. The larger the ring and the stronger the magnetic field, the higher the energy to which the particles can be accelerated. Quite a number of proton accelerators were built in the second half of the twentieth century, in the United States, in Western Europe and in Japan as well as in the former Soviet Union. Let us give special mention to the "Bevatron" in Berkeley, CA (built in 1954, reaching 6 GeV energy); the "Alternative Gradient Synchrotron", AGS, in Brookhaven, NY

(1960), 30 GeV; the PS accelerator at the European Center for Nuclear Research, CERN, in Geneva, Switzerland (1959), 25 GeV. Later on, there followed the accelerator at Fermilab outside Chicago (1972), starting at 200 GeV and raised to 400 GeV at a later time; and the SPS accelerator at CERN with a 26.7 km circumference, which reached 400 GeV. Significantly, the acceleration in most of these projects was not confined to one ring. It turned out to be advantageous to arrange a cascade of accelerating rings — such that the protons that were first boosted in energy by the PS ring, only to be subjected to final acceleration in CERN's SPS ring.

There was a significant leap forward for accelerator technology when superconducting magnets became available for the necessary magnetic field requirements. In a superconductor, the electric current finds no resistance, so there is no energy loss due to heating up conductors. But not just that this fact also permits much stronger currents in the conductors, but this, in turn, makes for stronger magnetic fields, maximally up to 10 Tesla. For that field strength, we need several thousand Amperes of current. Just for comparison: the Earth's magnetic field has a strength of only 1/20,000 Tesla.

Fermilab, in 1983, was able to accelerate protons to 900 GeV, just by using superconducting magnets. The Large Hadron Collider, LHC, which is being assembled at CERN, also uses superconducting magnets (of field strength 8.4 Tesla); in this way, it reaches 7 TeV (or 7,000 GeV) for both colliding beams. The LHC ring is being assembled in the same subterranean tunnel that hosted the LEP accelerator before.

Once the particles have been accelerated to velocities close to that of light, it is up to high energy physicists to proceed with their investigations. The particles are extracted from the accelerated ring, then to be directed at a "target", a piece of well-defined matter. The ensuing collisions permit the researchers to start their real work. The collisions of the high energy projectiles with an atomic nucleus happen in minuscule time windows — defined by the time it takes the proton to traverse the nucleus in about 10^{24} seconds, i.e., with a velocity close to that of light. It is impossible to follow the details of the collision, but its "final state" results can be observed. We know which particles collided, and we can observe the "final state" — what particles with what energies, in what directions.

It is almost like dealing with a traffic accident that happened without direct witnesses: the police investigators have to try and reconstruct what happened, from the facts at their disposal (damage to the vehicles and their final location, etc.). They can rely on the laws of classical mechanics being applicable to the traffic accident. In particle physics collisions, things are a good deal more complicated. The applicable laws of Nature do determine all details of what happens here; and we may or may not have a thorough knowledge of these laws — or may be only an approximate one, or none at all. One collision will not help us to learn more about these laws. But once we observe hundreds, or many millions of such events, we may well find out a great deal about the laws of microphysics.

Physicists have been able to improve steadily on the "detectors" of particle properties. Decades ago, they had just cloud chambers and scintillation counters. Then came bubble chambers which record particle tracks in fluids. Another technique to record particle tracks is the use of parallel metal plates that exchange small sparks between their different voltages once a charged particle passes between them: the so-called spark chambers. Today, electronic gadgets of considerable sophistication are mostly being applied. Their clear advantage is that they can feed electronic signals from detector elements immediately into computers that exploit the information they recorded. Obviously, this simplifies the analysis of particle collision events. On the distaff side, we may mind the fact that any kind of direct observation is now excluded, so that particle physics has largely lost its appealing feature of direct observability.

In particle collisions, just as in our everyday experiences, there is the well-known rule that momentum is conserved. A proton that comes with 1,000 GeV energy to hit another proton at rest has its own momentum of 1,000 GeV/c, but the proton at rest has none. After the collision, the overall momentum will be the same, 1,000 GeV/c. The sum of all final state particles must add up to those 1,000 GeV/c. What that practically means is this: after the collision, all participating particles exit in the direction of the incoming protons. Now recall that the principal aim of particle collisions is the transformation of a maximum of energy to mass — mostly that of new particles. It would be best if all of the energy could, at least in principle, be transformed into mass: in our example, that would be 1,000 GeV where we quantify the mass in units of energy. But the need to conserve momentum does not permit that; only a small fraction of the available energy can be transformed into mass — maximally it will be 42 GeV.

These conditions change if we collide the accelerated proton not with a proton at rest, but rather with another accelerated proton with the same energy but with the opposite momentum; that means the sum of the momenta will be zero. This case leads to a much more powerful collision: there is now a chance to come up with a total mass of 2,000 GeV, a great deal more than in the case of the target at rest. Compare this with a traffic collision — damages to the colliding car when it hits a car at rest are very much smaller than if it crashes into a car moving in the opposite direction. Frontal collisions with moving targets, to be sure, are much more difficult to set up than those with targets at rest. In the first case, we need two different particle beams; and to configure them such as to collide frontally with each other. Not only that: think of how easy it is to take a gun and shoot a bullet into a tree trunk. But now try and use it to hit another bullet that is coming your way. In particle physics it is not just the total energy that counts, it is also how often you manage to make a hit. We call that the luminosity.

Notwithstanding a lot of technical problems, it has been possible to achieve frontal particle collisions in flight. The machines where we manage that are called colliders. The first such device was taken into operation at CERN in 1972 and was called the Intersecting Storage Ring, or, for short, the ISR. Two beam pipes were mounted in parallel; then, 30 GeV protons were injected, and the beams were made to collide at specific points, yielding a total of 60 GeV for every two-proton collision. It took a lot of initial troubles until CERN physicists finally reached collision rates of almost 60 million per second, for the production of new particles.

At SLAC, their California colleagues built a relatively inexpensive collider in the early 1970s, called SPEAR. The principle was quite simple: electrons and positrons were injected into a magnetic ring configuration, such as to run in opposite directions. Positrons with their positive electric charge will run in precisely the opposite direction of the electrons in the appropriately configured magnetic field. Both leptons share the same evacuated rings as long as we make sure that they have the same energy. Once they collide, they annihilate each other, and all of their energy is available for the mass of new particle creation. Clearly, electron–positron collisions provide an ideal source for the investigation of new phenomena.

In fact, a series of significant discoveries were made at the SPEAR collider, to be followed soon by the DORIS ring at the DESY Laboratory in Hamburg, Germany. They contributed significantly to the build-up of what later came to be known as the Standard Model of Particle Physics. We will never forget the weekend of November 9, 1974: all the available energy of electrons and positrons in the SPEAR ring was transformed into one new heavy particle — which was later given the double name J/ψ, and which was to play an important part in the further development of particle physics.

In the years following this epochal discovery, a number of further electron–positron machines were built: there was PETRA at the DESY Laboratory in Hamburg, PEP at Stanford, and TRISTAN in Tsukuba, Japan. In the early 1990s, the LEP electron–positron collider permitted the first precision experiments at very high energies. The circumference of this machine was 26.7 km; the resulting research program turned out to be of the greatest importance for the consolidation of what we define today as the Standard Model of Particle Physics. Soon after LEP's seminal output began, similar energies were reached at the Stanford Linear Collider, the SLC. It utilized the linear accelerator to propel both electrons and positrons to about 50 GeV in the same beam line. They were then made to collide by the use of very powerful magnetic fields which managed to point them at each other.

The next pivotal progress was made, still in the 1990s, at the German DESY Lab, with the inauguration of the HERA accelerator. This machine collided electrons that had been accelerated to 30 GeV, head-on with protons of 800 GeV. This novel configuration permitted a very precise analysis of the inner structure of the proton – down to distances of 10^{-16} cm, or to a one-thousandth of the proton size.

When we accelerate electrons or positrons in circular accelerators, we have the disadvantage that those rings soon have to become very large. This is due to the electrons' rest mass, which causes their electromagnetic field to "protest" every change of their direction of motion by "making itself independent" and radiating off by itself. The ensuing radiation, called "synchrotron radiation", causes a ring accelerator to act as a steady source of this radiation — causing a steady loss of radiation energy. The closer we manage to get the electrons to the speed of light, the stronger the synchrotron radiation, so that, ultimately, no energy is left for further particle acceleration.

This problem does not exist when the rapidly moving electrons follow a straight line. That is why the linear acceleration of electrons and positrons is vastly preferable if we want to reach several hundred GeV. Consequently, there are now plans to create a large linear accelerator project as a global enterprise. Some 250 to 500 GeV are planned for each of the beams that will then be made to collide. A "Linear Collider" of this order of magnitude must be seen as an ideal tool for testing prevalent theories of particles and their interactions under extreme conditions, and for pointing our way beyond their realm of validity.

One decisive step in a new direction was made at CERN in the 1980s, when protons and antiprotons were caused to collide head-on at high energies. Antiprotons do not exist in the matter around us; but they can be produced in particle collisions and be made into beams which, in turn, can be accelerated just like protons. To be sure, the technical demands to produce a clean monochromatic antiproton beam, i.e., a beam where all particles are antiprotons of one given energy are enormous. It took years of intense efforts to achieve this goal.

Similarly to electron–positron machines, it is possible to make protons and antiprotons circulate in one and the same circular beam pipe in opposite directions if and only if, they share the same energy. As far back as in 1981, proton–antiproton collisions were initiated at CERN; each particle reached an energy of 400 GeV. The investment in this technical tour-de-force soon paid off. Soon after the first observation of "W- and Z-bosons", the mediators of the weak nuclear interactions were recorded. These are the particles that are responsible for the radioactive decay of the neutron, among many other phenomena. As of the early 1990s, the Fermilab laboratories outside Chicago took over the leadership in the investigation of proton–antiproton collisions. Just like it had been done at CERN, the large accelerator ring at Fermilab was reconfigured so as to make the investigation of proton–antiproton collisions possible.

In the above, we discussed electron and positron acceleration and the problems caused by synchrotron radiation. Similar problems basically also exist in the proton–antiproton rings, but they pose no practical problem. The heavy masses of these particles do not emit similar radiation. We can accelerate protons and antiprotons in circular machines to thousands of GeV without running the risk of having them emit noticeable radiation losses. Sometime in the years 2006–2007, the LHC accelerator at CERN will go into operation. It will accelerate protons to 14 TeV in the circular tunnel of the LEP accelerator, making use of two vacuum tubes just like the ISR we discussed before, making frontal collisions possible. This collider will be the first to penetrate the TeV range of our beam energy scale. We are not able to tell what new phenomena we are going to observe there, although there is no dearth of speculation. There is hardly any particle physicist who doubts that there will be new surprises awaiting us at energies that become observable at LHC energies.

Let's not finish this chapter without adding a few words on the methods of observing and recording particle interactions in these experiments. Up until about 1970, observations of particles were mostly made with the use of cloud chambers and bubble chambers, which we will not describe here except to say that electrically charged particles left little tracks when passing through them — tracks that were visible and could be photographed. In the early 1970s, it became possible to record the tracks electronically. While their observation became less picturesque this way, this transition opened the door to immediate computer analysis. Also, this transition made for greater flexibility and permitted the registration of rapid sequences of observation. It also opened the door to devising triggers for certain processes among all the phenomena that occur. One important feature that we cannot properly cover in this text is the development of detector technology, which is a vital component of every new discovery.

Chapter 5

Quarks Inside Atomic Nuclei

Let us recall that Rutherford used alpha particles to investigate the structure of atoms. When we consider the investigation of atomic nuclei and other particles, alpha particles are completely useless. For one thing, they themselves are nothing but helium nuclei; and their energies of just a few MeV are quite inadequate for any further investigation. That might sound self-contradictory, but it is a simple consequence of quantum theory: the smaller the structures we want to investigate, the larger the energy requests on the probes we need. The uncertainty relation tells us that a high momentum gives us access to tiny structures, and vice versa. It is easy to calculate that we can use particles of just a few MeV energy — say, alpha particles as in Rutherford's experiment — to observe structures as minuscule as 10^{-12} cm. And that's the best resolution we have access to with "slow particle" probes. Once we need to resolve smaller structures, we have to resort to higher energies. And that is why particle physics experiments need the construction of accelerators that can propel electrons, protons and other charged elementary particles to high energies.

To probe for new structures, electrons are particularly useful. Why? Because, as best as we know, they have no structure of their own, they are "point-like" — unlike the nucleus. When it was first noticed that atomic nuclei have radii of order 10^{-12} to 10^{-13} cm, it was not clear whether this is due to the strong interaction forces between their constituents or maybe simply to the extent of the nucleons. This question was resolved in the 1950s when protons and neutrons were probed by high energy electrons: after all, the size of nucleons can be "measured" by scattering electrons off them. An electron that passes close by a nucleon will suffer a change in its incident direction; the extent of this change depends on the distribution of electron charges inside the nucleons. In this fashion, it was noticed that a proton does not have a point-like charge distribution like the electron — instead, its charge is distributed inside a sphere of about 10^{-13} cm diameter. Similarly, it turned out that a neutron, notwithstanding its net charge of zero, has a charge distribution in its structure — but positive and negative charges compensate each other. It was first noticed that electric charges occupy a neutron's inside; but to what particles those charges belong, that was unknown. There was a vague agreement that electric charges are characterized by a continuous distribution, a kind of electrical porridge.

It took many years until, in 1964, an important step in the direction of a comprehension of the charge structure of nucleons was made: quite independently of each other. Murray Gell-Mann of the California Institute of Technology and the US physicist George Zweig, then at CERN, suggested that the nucleons are composites of electrically charged particles of spin 1/2, to be called "quarks". The nomenclature, introduced by Gell-Mann, is a verbal concoction first popping up in James Joyce's novel "Finnegan's Wake". It must be mentioned right away that it was not due to electron–nucleon scattering results that Gell-Mann and Zweig chose this name for the hypothetical particle, but rather the fact that such a hypothesis was able to explain a number of new phenomena which were clearly connected to recently discovered particles and their symmetry properties.

To build up protons and neutrons, we need two different quarks which were given the symbols u and d (for "up" and "down"); the proton consists of two u-quarks and one d-quark. For short, we write $p = (uud)$. Analogously, the neutron inverts the composition: $n = (ddu)$. The antiproton contains two anti-u-quarks and one anti-d-quark, the antineutron two anti-d-quark and one anti-u-quark.

A pivotal point of interest concerns the electric charges of the quarks: the u-quark's charge is non-integral if measured in terms of the proton charge, it is 2/3. That of the d-quark is one third that of the electron, or $-1/3$. Consequently, the charges of the antiquarks are 2/3 and +1/3. The charges of the two particles making up all nuclei, the protons and neutrons, result simply from adding up the quark's electric charge. That means a proton charge of $2/3 + 2/3 - 1/3 = 1$, and a neutron charge of $2/3 - 1/3 - 1/3 = 0$. So, there are charged constituents inside the neutrons — but for the outside they cancel out to zero. The odd one-third amounts of the quarks' electric charges must be seen as their most surprising properties; and that fact originally elicited a good deal of criticism. Zweig, in fact, had trouble getting his work on this subject published.

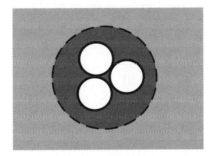

Fig. 5.1.   Three quarks inside the proton.

Given the uncommon electrical charges of the quarks, some physicists originally did not feel they were the actual building blocks of nuclear particles; rather, they viewed them as abstract mathematical symbols that helped describe the particles and their symmetry properties. Nobody considered the possibility of observing quarks directly in an experiment.

When a rapidly moving electron collides with a point charge — say, with another electron — the usual result is a scattering of the electron; it will change its direction. This implies a momentum transfer from one particle to the other. This concept makes sense only if the charge of the target particle is strongly concentrated. This might be seen as a parallel to Rutherford's experiment with alpha particles which taught us that the positive electric charge is concentrated in the atomic nucleus. In 1967, experiments in the new Linear Accelerator Laboratory SLAC in Stanford, CA, started to investigate nuclear structure along these lines. The strongly accelerated electrons were directed at atomic nuclei, and the resulting change in their direction contained information on the charge distribution inside these "targets". Nobody expected the results of these SLAC experiments to come up with answers as simple as they turned out.

To their considerable surprise, physicists noticed that, once in a while, an electron was strongly deviated from its incoming direction, just like in Rutherford's experiment with alpha particles. The implication was that electrons, while traversing nuclei, will sometimes hit on a point-like charged object. When those data underwent a more refined analysis through the following years, it became possible to draw conclusions on the electric charges of the scattering objects inside the nuclear material — and lo and behold, those charges turned out to be $2/3$ and $-1/3$! Apparently, those quarks were not just hypothetical objects.

During the experiments at SLAC, the incident electrons, traveling at the speed of light, were directed at nuclear matter at rest. To get a hold of a viable interpretation of the data, it is helpful if we try to adopt the point of view of a fast-traveling observer moving with the incident electron. We are not changing the process in this formulation; we are just adopting a different point of observation. Compare this to that of looking at the collision of two cars — either from the vantage point of an observer at rest, or from that of a passenger in a railroad train that happens to pass by during the collision. To wit: the result of the collision, the damage to the two cars, does not at all depend on the observer.

From this vantage point, the moving observer sees the nuclear particles and the electron collide frontally at high velocity. So what happens to the proton when an electron collides with a quark inside the proton? Only one of the three quarks inside will be hit, and so the other two quarks move on as though nothing had happened; but the third quark, hit by the electron, will move in a different direction, far away from the other quarks. The colliding electron will be sent off in the opposite direction. So we expect the proton to be split up into its quarks. But far from it: once we perform the experiment, we notice that, opposing the exiting electron, a beam of particles will leave the interaction point. Most of these will be what we call mesons, mixtures of quarks and anti-quarks, which we will describe below. Similarly, the two quarks that did change their direction of motion will not behave just like bare quarks: they transform into a beam of particles that usually contains a proton or a neutron. That is to say: we are not really dealing with a dissection of the proton into its constituent particles.

This is where our result becomes incomparable to the Rutherford scattering experiment we have been quoting: when an alpha particle collides with an atomic nucleus, that nucleus will be ejected from the target material, simply because the forces inside the atom are relatively weak. But the nuclear particles have a different make-up: the forces between the quarks are obviously of the strength that does not permit the ejection of a quark due to the collision with the electron. But if these forces have such signal strength, why is it that the quarks inside the nucleus behave like point particles, i.e., like particles that are not subject to any forces outside the electrical one?

A closer look at the experimental results put another riddle to the physicists involved. When an electron of large momentum is scattered from its incident path when colliding with a nuclear particle, there are two quantities that carry relevant information on the scattering: the angle of scattering (say, 20°) and the energy loss it suffers in the collision (say, it is incident with 20 GeV, then exits with just 12 GeV). The final-state angles and energies are straightforwardly measurable, and they carry information about the momenta of the quarks in a rapidly moving nucleon. Say, a proton, consisting of three quarks, moves rapidly with an energy of 18 GeV, we might well assume like a composite of three quarks, each of which has an energy of 6 GeV, i.e., one third of the proton energy. In the same vein, each quark ought to carry a momentum one third of the proton momentum.

But the experimental results do not bear this out: the quarks do not carry roughly one third of the proton momentum each; rather, they show an interesting momentum distribution. Sometimes a quark actually carries about one third of the full momentum, but quite frequently it carries a good deal less — say, only about 1/10 of the initial momentum. This observation is quite significant for the function of the quarks; it tells us a lot about the forces acting between them, as we will discuss. The physicists were also stunned by the fact that, on the average, the fractional momentum each quark carried was well less than expected. After a good series of quark momentum distribution measurements, it turned out that the summing up of the quark momenta did not yield the full momentum of the nucleon — rather, those momenta added up to no more than about 50%.

That means: the quarks carry only about one-half of the momentum of a rapidly moving nucleon. But where is the rest of the momentum? Are there more constituents to the nucleon that, for some reason, have not been observable in the electron–nucleon scattering experiments? That would imply the following conclusion: the experiments observe only electrically charged constituents, so that these additional constituents would have to be electrically neutral. They do not contribute to the overall charge, but are part of the nucleon momentum. The implication was that these neutral particles or quanta had to be part of the hitherto unfathomable forces between the quarks. They make up the glue that "sticks three quarks together" to make up a nucleon, a proton or a neutron. The name that was given to these particles/quanta describes their function: they were called gluons (= glue quanta).

So, altogether, the nucleons consist of two types of quarks: u and d. For the designation of the different types of quarks, the term "flavor" was chosen. There are the quark flavors u and d (and there are more flavors that we have not encountered yet). Up to this point, these two quarks, u and d, are the elementary constituents of atomic nuclei — where we have to explain our choice of the term "constituents". Usually, this term implies that a nucleon should be divisible into these parts: but to this day it has not been possible to break a nucleon up into quarks — and there is no prospect that this breakup will ever be feasible. When speaking of the nucleons' substructure, it appears that we have arrived at a level where our notions of further divisibility, based on observations that do no longer apply here, do not make anymore.

If we now add the electron as a building block to the outer layer of atoms, we may well say that "normal matter" throughout the Universe consists of u quarks, d quarks, and electrons. On this level, particle physics as the science describing these constituents appears relatively straightforward. But Nature has not stopped at these elementary objects. The world of elementary particles is much more complex than the study of the structure of "normal" matter would make us suspect. Beyond those u and d quarks and electrons, there is another world of unstable particles which we can penetrate only with the assistance of particle accelerators.

The first steps to penetrate this new world which we aim to penetrate soon in our present inquiry, were made in 1937: there were no accelerators available; rather, the decisive instrument was simply a cloud chamber that permits the observer to see "tracks" of charged particles from cosmic radiation — an instrument that looks pretty antiquated today. We know that the upper layers of the Earth's atmosphere is constantly being bombarded by rapidly moving particles that arrive somehow from the depth of the Universe. Most of these are protons or light atomic nuclei, such as deuterons or alpha particles. When these collide with atomic nuclei in the atmosphere, short-lived charged particles are being created that decay after no more than a few millionth of a second, and there will be an electron among the decay products. Notwithstanding their short lifetime, these particles often manage to penetrate our atmosphere to reach the surface of the Earth. This is due to the theory of relativity which tells us that time moves more slowly for rapidly moving particles when compared with particles at rest.

These new particles, called muons (after the Greek letter μ, pronounced mu), have a mass of about 107.5 MeV, making them about 200 times as heavy as an electron. Further experiments in the 1940s showed that these particles, notwithstanding their considerable mass, are "point particles" like the electrons — we might call them the heavy cousins of the electrons, the electric charges of which they share. Their only difference from the electrons is the considerably heavier mass and the fact that they are not stable. For the structure of "normal matter", these heavy cousins of the electrons, for short the "muons", make no difference. They appear to be a useless addition to the particle zoo. The US physicist Isidore Rabi is quoted as having asked: "Who ordered that?" Nobody to this day has been able to answer Rabi's question. And yet, there are indications that the muons have their special part in the symphony of sub-nuclear particles, as we will see.

Chapter 6

Quantum Electrodynamics

When two objects exert an influence on each other, this fact shows that they must share a bond which, however, can be realized in different ways. There is usually a force they exert on each other. Take, for example, the Earth's gravitational attraction of an apple dangling from a tree: at some point, the apple will fall to the ground. It turns out that, to complete this description, not just the Earth as a massive object exerts a force on the apple, but so does the apple attract the Earth. We are dealing with a reciprocal phenomenon — as is always the case when a force acts between two objects. That is why physicists often do not describe such phenomena in terms of forces, but rather in terms of interactions — in our case, of a gravitational interaction. Gravitation is not the only interaction we observe in Nature as a macroscopic phenomenon, although it is the most ubiquitous one. But we also frequently observe, in our daily lives, electrical forces of interaction — objects of the same electric charge repelling each other, while those of opposite charge attract each other. Our everyday life also observes magnetic interactions, such as the movement of a compass needle that points to the North, or, in a different vein, when we witness a patient's subjection to "nuclear spin tomography" in a hospital.

Up till the early 19th century, scientists categorized electrical and magnetic interactions separately. They were seen as independent phenomena. This view changed as soon as it was found out that electric currents or the movement of electrically charged objects generate magnetic fields of force. It was then noticed that electric currents can be caused by rapidly changing of the magnetic forces. This effect is being exploited today as power stations generate electric currents.

The mutual relations that have been observed between electric and magnetic phenomena caused the assumption that there must be a link connecting them. This became clear when the British physicist Michael Faraday introduced a new concept, that of electric and magnetic "fields". In this way, he changed the perspective of scientists concerning the forces observed. Up to that time, the concept had been that electric, magnetic, and even gravitational forces were phenomena that act over some given distance — e.g., the Earth attracts the moon because it exerts a force over some 300,000 km.

Faraday, on the other hand, figured the electric forces to be due to a field that originates from electrically charged objects and fills the surrounding space with lines of force: the two of them attract each other because the space between them is filled with these field lines. They exert an influence on each other, and that makes us understand the mutual relations of the electric and magnetic phenomena. Hence the general talk about "the electromagnetic field". In addition, it turned out that Einstein's theory of relativity causes magnetic and electric fields to be seen as one phenomenon: a magnetic field that is analyzed by a rapidly moving observer is not just a magnetic field; it turns out to be a mixture of an electric and a magnetic field.

Later on it turned out that the notion of a "field" is of the greatest consequence in physics. Were it not for those fields, a quantitative description of many natural phenomena would be impossible. Oddly enough, they are a concept that appears strange to every layman for starters: our perceptive senses do not notice them directly. We humans have a sense that makes us appreciate sound waves but not electromagnetic waves. Modern theories that help to delineate the "behavior" of elementary particles are field theories, without any exception. Every physicist, every engineer, knows these days that fields are a reality: just like material objects, fields carry such phenomena as energy and momentum.

Those fields need not all be tied to massive bodies — they may also have a "life" of their own. The equations that describe the properties of electromagnetic fields were formulated in 1861 by the Scottish physicist James Clark Maxwell, and they bear his name to this day. Next to Newton's laws of mechanics, Maxwell's equations are the theoretical columns that carry a good part of modern technology, one of the significant implication of these equations is the fact that changes of electric and magnetic fields propagate in space with the velocity of light. That is by no means a coincidence: light, after all, is nothing but an electromagnetic phenomenon. Light waves are electromagnetic waves that are perceived by our eyes simply because they are receptive to their signals, but not to electromagnetic waves of different wavelengths. Take Roentgen's radiation — it has electromagnetic waves with wavelengths that are shorter than normal light, just as radio waves have longer wavelengths — "short waves", for instance, have wavelengths of some 10 meters.

When we talked about the quantum aspects of light, we mentioned that electromagnetic waves have quantum properties. When such a wave spreads in space, its energy is not transmitted with full continuity as classical wave phenomena would make us expect; rather, there are small energy clusters, which we call photons (or light quanta). Maxwell's equations of electromagnetism describe photon dynamics when we interpret them in the framework of quantum physics, establishing the tight links of the two aspects. In this way, we establish the theory of quantum electrodynamics, usually called QED for short, as it was developed in the 1930s. Pioneers in this effort were Werner Heisenberg and his Viennese colleague Wolfgang Pauli who contributed vital parts to this development, which was based on the investigations of the British theorist Paul Dirac. We note particularly that quantum electrodynamics combines the elements of quantum mechanics, of field theory, and of the theory of relativity.

In this framework of electrodynamic theory we find an important new interpretation of electromagnetic forces. Let us take, for instance, two electrons scattering off each other: the two particles move towards one another, pass by each other, then move apart. Electromagnetic forces make them repel each other so that their direction of motion changes: they are "scattered" and move away in different directions; their "scattering angle" depends on the details of their tracks. How do we interpret this scattering process in the theoretical framework of quantum electrodynamics, where the electromagnetic field has quantum properties, and where the quantized energy is carried by photon "particles".

An electron is surrounded by an electromagnetic field which can be described in terms of photons. A rapidly moving electron can best be imagined as a charged massive object which moves across space surrounded by a cloud of photons. When two such electrons pass by each other, those photon clouds get mixed up. Since a photon is not individually tagged, there will be photon exchanges. A photon from the first electron's cloud will make part of the other electron's, and vice versa. Since each of the photons carries energy and momentum, these exchanges cause change in the individual momenta, and thereby of the flight direction of both electrons — part of the action of the electromagnetic force. Since we are talking about two electrons, this force is repulsive. If, on the other hand, we consider an electron interacting with a positron in an analogous way, there will be a photon exchange, but this time the force is attractive.

A key element of quantum electrodynamics is the fact that it implies: the forces between charged particles are carried by the photons: those "quasi"-particles that are in reality the "particles" of light. Let us note, however, that there is an important difference between the "particles of light" and the photons that mediate the electromagnetic forces. The first of these can be seen as "free" or "real" photons, since they are independent and do not just make up the "cloud of charge" of a charged particle. In their case, it is important to note that both their energy and their momentum, measured in units of eV, are precisely the same. A photon may come out from a nuclear reaction and may have the energy of 1 MeV, but its momentum points in one direction and also measures 1 MeV. Particles with this property have no mass since particles of mass M have, even if at rest, the energy $E = mc^2$, but no momentum. They therefore do not qualify as massless particles when at rest, because their energy does not equal their momentum. The photons that are responsible for the transfer of force between charged particles may, due to the uncertainty relation, dispose of arbitrary energies and momenta. That is what lets us call them "virtual particles". So, the electric attraction between two oppositely charged spheres is due to the exchange of virtual photons.

The strength of the force depends on that of the interaction between the quanta of that force, i.e., of the photons, and of the charged particles. Notwithstanding the fact that electric attraction or repulsion may exist over a certain distance, the actual interaction is the contact between electron, positron and photon. In this case, we talk of a local interaction, because the contact between a charged particle and a photon is point-like. The German physicist Arnold Sommerfeld noticed in 1916 that the interaction strength between photons and electrons can be described by a pure number, which he called the "fine structure constant" and to which he assigned the symbol α (the Greek letter alpha). The name he gave means that it tells us about the detailed structure of atomic energy levels. It expresses that this number carries information on what subsequently became clear in the development of quantum electrodynamics — the combination of relativity, quantum theory, and electrodynamics: the quantity α is defined as e^2/hc, with e the symbol for electric charge. So, it contains the electric charge e, the constant of quantum theory, h, and the velocity of light, c, which serves as the basic constant of the theory of relativity.

This constant "alpha" is just a number, it carries no dimension (say, m or cm or sec). It has to be determined by experiment. Its value is well known these days, but for the present purpose we might limit the precision to one part in a million and call its number $\alpha = 1/137.036$: this is a small number, smaller than 0.01 — and its reciprocal value is almost an integral number, 137.

Let us emphasize that this number is the most prominent number in all of the natural sciences. Ever since its first introduction it has caused a lot of speculation. After all, α gives the strength of the electromagnetic interaction, which gives it fundamental importance for all the natural sciences and for all technology. If it had a different numerical value, that would change many things in our daily

lives. After all, the structure of atoms and molecules is based on it.
If it were just slightly different, many complex molecular structures
would not exist as stable systems, and that fact, in turn, would have
consequences in biology that we cannot even determine. Quite obviously, a theoretical determination of the numerical value of α would
signify great progress in our understanding of fundamental interactions. Many physicists have tried to find it, but without significant
success to this day. Richard Feynman, the theory wizard of Caltech
in Pasadena, once suggested that every one of his theory colleagues
should write on the blackboard in his office: 137 — how shamefully
little we understand!

As we mentioned above, electric and magnetic forces are phenomena that are due to a local interaction between charged particles and
photons. When two electrons repel each other, there are actually two
electromagnetic interactions: first the emission of a virtual photon in
one point — which we call a vertex, and then the absorption of the
photon in another vertex. Such sequences are described in the form
of diagrams (see e.g. Fig. 6.1) that have been named after Richard
Feynman, who first used them. The two elementary interactions are
marked by the quantity of electric charges, written as the letter e.
The force is then proportional to e^2 hence the appearance of it in the
"fine structure constant", which we introduced earlier. If we were
not dealing with electrons but, say, with the repulsion of two alpha
particles — which have a charge of $2e$ each instead of just e, the
repulsive force would be four times as strong.

The simple fact that α is a fairly small number has remarkable
consequences when we describe processes in quantum electrodynamics. For one thing, it means that electromagnetic interactions are
fairly weak (when compared with the strong nuclear interactions).
When an electron interacts with matter — say, by "hitting" an atom,
nothing dramatic is liable to happen. Most probably, the electron
will just be deviated slightly in its track. In just about 1% of those
incidents will there be a stronger deviation in its path, because the
likelihood for that process is of the order of magnitude of alpha.

The small value of alpha also permits us to calculate quantum
corrections to the basic processes of QED — a procedure which we

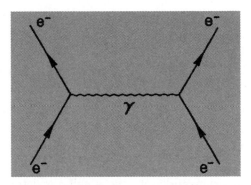

Fig. 6.1. The repulsion of two electrons is described in quantum electrodynamics by the exchange of a virtual photon between the two electrons, as illustrated in the Feynman diagram. The local interaction happens at two points, which we call vertices, the virtual photon moves from vertex to vertex, transmitting the electromagnetic force in this fashion.

call perturbation theory. It deals with what we call processes of higher order — like the exchange of not just one but two photons between two electrons. In this case, the basic electromagnetic interaction has to be entered four times. That implies that the strength of this process is proportional to the fourth power of e, or to α^2. Now, recalling that $\alpha = 0.0073$, we now have an effect of order α^2 or of 0.00005. Further quantum corrections are proportional to α^3, i.e., of 0.0000004, and so on. Those quantum corrections, tiny as they are, can be experimentally verified, and we see that theory and experiment agree very well. In fact, they agree to one part in a billion. Without the slightest exaggeration we can say that QED (short for quantum electrodynamics) is the most successful of all our theories dealing with elementary particles and the forces between them. Not only that, QED also "paved the way" for a successful quantum field theory. Today's theories of particle physics have been fashioned after the example of QED.

In addition to the electrons, photons and antiparticles, we include the positrons as they turn out to be important in our case. Without doing so, there would be no way to formulate QED without internal contradictions. The existence of antiparticles is needed for a successful quantum theory of electromagnetic interactions. Paul Dirac noticed that as early as 1932, before the first observation of positrons.

He tried, in the late 1920s, to combine the newly developed quantum theory with the theory of relativity. In the process, he wrote down an equation which turned out to be essential for the further development of QED and of particle physics; it has been called, in his honor, the Dirac equation.

His equation describes particles with spin 1/2, such as we would naturally expect in the case of electrons. That made the specific value for the electrons' spin a necessary consequence of the theory of relativity and of quantum mechanics. He then found out that his relevant equation would not give a self-consistent description of the quantum properties of the electron if the existence of an anti-electron with the opposite charge was not assumed. In this way, the positron was created, for starters as a theoretical construction. The experimental proof of this assumption followed immediately — positrons were noticed by Carl Anderson at Caltech in cosmic ray observations.

When we calculate processes in quantum electrodynamics, it is always necessary to include positrons. The theory assumes that the particles involved: electrons, positrons, photons, have no inner structure. An electron is simply a massive point particle which interacts electromagnetically. But in reality things are not quite that simple. Quantum theory tells us that the uncertainty relations imply: empty space is not quite empty, but replete with virtual pairs of electrons and positrons, i.e., with particles that exist only at tiny distances and over very short periods of time. The vacuum hosts a steady set of pair creations and pair annihilations. Viewed from close up, the vacuum is not a quiet abode, the way classical physics would have it. The smaller the distance we explore, the more violent the activities of those virtual particles. The closer we look the vacuum becomes a witches' kettle rather than empty space. We might compare it with the ocean: from a plane up in the sky, we see way below us an infinite wide plane water surface — to all appearances the image of empty, perfectly two-dimensional space. But the lower our plane flies, the more this image changes. First, we see slight surface crinkles; then, as we get closer, we notice that there are waves which may rise and collapse. The same is true with "empty space", the vacuum, at a

safe distance, is just that quiet volume which our senses perceive —
but then, the better we resolve the details of space and time, things
really look different. Virtual electron–positron pairs are constantly
being produced and annihilated. In terms of microphysics, there is
a steady hectic activity, the dance of virtual particles which remains
macroscopically invisible — its effects average out to nothing.

So, there are no macroscopic consequences to the activities of the
virtual particles — but the immediate surroundings of, say, an electron are influenced by them. We might conjure up this image: suppose we put an electron in some well-defined point in space. Being
negatively charged, it will repel the virtual electrons in its neigh-
borhood while attracting the virtual positrons. We call this process
vacuum polarization. In the immediate vicinity of an electron, its
charge is partially screened off by virtual positrons. If we try to look
at an electron from the outside but nearby, we will not see a point-
like particle. What we see is the electron together with the cloud of
virtual particles that surrounds it. This is often called a "physical"
electron to differentiate it from a "naked" electron, without its "polarization cloud". The charge of the latter has to be larger than that
of the "physical electron".

This implies that, as a consequence of quantum theory, a "point-
like" electron is not as point-like after all. Seen from some distance,
it may well appear so. But when we penetrate to distances of, say, a
hundredth of an atomic diameter, the effects of vacuum polarization
became noticeable, and that leads to effects in the quantum electro-
dynamical context. If we were to calculate the charge of a "naked"
electron when compared with the measured charge of the physical
electron, we wind up with a senseless result — it is infinite!

This finding is not the only disagreeable surprise QED gives us:
Let us look at the mass of the electron, the equivalence of energy and
mass causes the electric field of an electron to add to its mass, after
all, the existence of the field must mean that there is a certain energy
density when we see that field. But again: if we calculate the contribution of that field to the mass, we again hit on a senseless result.
That is not even that strange — because our theory says the electron has no inner structure at all, it is strictly pointless. That must

imply that the electric field becomes very strong at tiny distances; a quantitative evaluation again shows that the field contributes an infinite amount to the mass. Again the assumption that the internal structure of the electron is "infinitely small", i.e., strictly point-like, leads to senseless infinities.

Maybe, we could say, the electron does have some structure, but that becomes noticeable only at very small "distances", say, at 10^{-18} cm. So it may have a non-zero radius after all. It is easy to show that this assumption avoids the painful infinities. But try as our theorists might, we have to date not been able to detect any experimental hint that this might be so. Instead, we have been able to come up with a limit on a non-zero radius for the electron, it has to be smaller than one-hundredth of the size of a nucleon.

It is a fair question whether Quantum Electrodynamics is a successful theory although it yields obviously abstruse results, as we saw. The "infinite" charge of the bare electron is not a serious problem, because this "naked particle" is not a physical fact — rather, it is a theoretical construction, a product of our mathematical thought — simply due to our artificial division of the electron into that central "naked" electron and its surroundings, a coating of virtual particles. This infinity can be absorbed by declaring that the charge of the bare particle equals the experimentally defected phenomenon, simply ignoring the pseudo-charge of the "naked particle". In the same vein, we equate the electron's mass with the measured value, paying no attention to the fact that a straightforward calculation yields a senseless result. This procedure is called, for short, renormalization.

It is then clear that we have a description of the process of quantum electrodynamics that is logically consistent. The infinities that pop up in the formal considerations do not affect the measurable physical quantities. Once, when Richard Feynman, one of the "inventors" of renormalization theory, talked about the fact that infinities that pop up in the theory actually cancel each other as long as we stick with measurable quantities, there was this interjection from his colleague Robert Oppenheimer: "the fact that a quantity has infinities does not necessarily imply that it is, in fact, zero". Still, it turned out that the irony in these words was not really called for.

Seen from today's vantage point, the pragmatism that is at the basis of renormalization theory is fully justified. Oppenheimer, Dick Feynman's skeptical colleague, was, of course, the director of the "atom bomb project" in the Second World War.

It is important to recall that access to the quantum physical descriptions of electrodynamic processes opened the door to some remarkable successes. A short time after Dirac put out his famous equation, the Dirac equation, he was able to show that in the framework of his access to electron phenomena, these very electrons had to possess magnetic properties that are described in terms of a specific magnetic moment — and indeed, Dirac's definition of the specific magnetic moment, that is so closely coupled to the electron spin, was found to be what experimentation verified. Decades later, physicists noted that there does exist a small difference between the experimental value and Dirac's prediction — experiment saw a "surplus" of order 0.1%; that constituted a challenge for theorists, who, indeed, solved this micro-riddle with bravura. It was shown that, strictly in the framework of QED, this tiny deviation is a consequence of renormalization, i.e., an effect of the virtual particles that surround the electron. A calculation of the magnetic moment shows that the infinities cancel out. The resulting effect is so simple that we can easily mention it here: the magnetic moment is larger by $\alpha/2\pi$, i.e., numerically, by about 0.1%. Today, the magnetic moment has been determined much more precisely, and to justify this theoretically, elaborate calculations of the quantum phenomena caused by the virtual particles have to be performed. The resulting agreement between theory and experiment is astonishing: it gives convincing proof that quantum electrodynamics indeed yields the correct theoretical description of the micro-physical process involved.

We still have not mentioned another quantum effect affecting the precise amount of electric charge. We mentioned that the charge of the electron is partially shielded by the cloud of virtual particles. Now, if we managed to take away a part of the charge cloud of virtual particles surrounding the electron, the shielding they provide becomes slightly smaller, increasing the electric charge slightly above the measured value: this would imply a tiny increase in the fine structure constant. A partial removal of the charge cloud around a physical electron cannot be directly effected, at least no more than instantaneously. But the effect can be tested differently: let an electron collide at high energy with another electron or positron, such that these particles get "very close" to each other — so close, indeed, that the shielding by the charge clouds is partially cancelled. In other words, during the collision, the particles act as though the effective value of α were slightly larger than the one measured in atomic physics. The experimental investigation was realized in the last decade of the 20th century at the LEP collider at CERN and at the SLC in Stanford, CA. It turns out that the measured value of the "fine structure constant" alpha is larger by 7% than the value mentioned before, just as expected in the framework of our theory. There could not have been a more impressive agreement between theory and experiment.

QED theory, originally conceived by Dirac, Heisenberg, Pauli, and other theoretical luminaries in order to describe phenomena of atomic physics, was astonishingly tenacious. Up to energies that exceeded the typical atomic applications by factors of a billion, its field equations turned out to be reliable guidelines for the description of basic processes. Those facts alone would have been sufficient to name the creation of QED as one of the intellectual pillars of the 20th century. But in reality, its importance goes well beyond this: it is the example after which today's theories of fundamental particles and fields were fashioned in the second half of that barely bygone century.

But before ending this chapter, let us mention one more important property of QED. When the German mathematician and theoretical physicist Herman Weyl took a close look at its equations in the 1930s, he noticed a new symmetry that had been automatically included without even having intentionally included it: it was given the name of gauge theory. The quantized field that describes electrons and positrons is, as the nomenclature goes, a complex field. That is, it is described in every point of space by a complex number. Such a number actually consists of two numbers that define its place in a particular type of plane, a complex plane. To define it fully, it is useful to describe it by its distance from the origin and by an angle, the so-called phase angle.

The Dirac equation has the property that it does not fix the phase angle of the electron field. It leaves that angle arbitrary. We can rotate without, in the process, changing anything about the electrons described by the field. We call this a gauge transformation. But it is important that the same rotation will be performed in every point of that space. If we turn the phase angle of the field by $20°$ in Paris, we have to do the same in Berlin. Were we to change this angle in Berlin by $10°$, in New York by $37°$, in Tokyo by $73°$, we would have a problem. It is perfectly alright to change the gauge angle of the field globally, but not just locally. That is why we may perform a global re-gauging of the field, but not a local one. And we mention that there is a global gauge symmetry, but not a local one.

Weyl was correct when he was uncomfortable with this form, and he tried to change the equation such that it should become possible to introduce an arbitrary rotation at every point. For an electron field that is not subject to any interaction, this is not possible, as we mentioned above; but it can well happen when the electron field has an interaction with an external electromagnetic field, as we can observe in Nature. The changes that pop up in the relevant equations when the phase angle rotates can be compensated by a change in the delineation of the electromagnetic field. This is what is called a re-gauging of the electromagnetic field. The two gauge procedures, the re-gauging of the electron field by a rotation of the phase angle, and the re-gauging of the electromagnetic field have, and this is of vital importance, to be fitted to interact precisely like two cog wheels in the gear unit of a car. That alone allows a force to be transmitted, in the automobile as in electromagnetism. That is when a local gauge symmetry has been established. This joint action of the two gauge changes is possible only when the quanta of the electromagnetic field, i.e., the photons, have spin 1 — and that is clearly the case. Were the photons scalars without spin, this joint action would be out of the question.

We have now established that the electromagnetic forces are due to this local gauge symmetry. The interacting system consisting of electron, positron, and photon therefore exhibits a particularly high degree of symmetry. In addition, it is easy to show that the local gauge symmetry enforces the masslessness of the photon. It would not be hard to change the equations of QED such as to give some non-zero mass value to the photon — but that would imply we would have to give up the local gauge symmetry. In addition, the symmetry implies that electric charge is a strictly constant quantity. It cannot be created or thrown away. If and when a "system" has been given charge, we can be sure that it will not change over the course of time, unless, of course, it establishes contact with another such system. Charge conservation and local gauge symmetry are very closely related.

We have now established that QED is a theory of the interaction of charged particles and of photons based on local gauge symmetry. That is why we call it a gauge theory — a term that was not created by Weyl, but only about four decades after his groundbreaking work.

When he discovered the local gauge symmetry of QED, this was by no means coincidental. Weyl followed a theory trend that had been established by Albert Einstein. In 1916, Einstein had published his theory of gravitation — the General Theory of Relativity. In this theory, there are also gauge transformations. They are different from those of QED, and for our present purposes there is no need to go into the details of these differences. But just as in the case of QED, the gauge transformations in Einstein's theory are closely intertwined with the relevant conservation laws. In the case of gravity, these are the laws of constant energy and momentum. We will show that the concept of local gauge symmetry has important repercussions well beyond the frame of QED phenomena: it is of basic importance for the description of the interactions of elementary particles.

Chapter 7

Quantum Chromodynamics

Using high-energy electrons from the Stanford Linear Accelerator Center (SLAC) in the late 1970s, physicists managed to get an X-ray picture of nucleons (protons and neutrons). What they saw was, mainly, an image of three electrically charged but structureless objects inside those particles — but their charges had the odd values 2/3 and −1/3. More detailed evaluation of those objects showed that those quarks have spin 1/2. Given that the spin of a particle is composed of the spins of its constituents, and that there is no need to pay attention to orbital angular momentum effects, it became clear that the proton spin is a combination of the three quark spins: two of the quark spins cancel each other, and that of the third quark determines the spin of the nucleon.

Five years before it was noticed that the quarks are "point particles" (i.e., have no inner structure) inside the nucleons, another odd property of the quarks had been noticed in the framework of the naive quark model of Gell-Mann and Zweig. It should be noted that their model had already confirmed the nucleons as three-quark structures. Electron–nucleon scattering experiments in the 1950s had already given signals of an excited state of the proton; it was short-lived, and, oddly enough, had an electric charge of 2 and spin 3/2. This object of mass around 1230 MeV was then called the Δ^{++} particle. In the framework of the quark model, it had to be made up of three u

quarks, the charges of which add up to 2. And if all three quark spins point in the same direction, they will add up to 3/2, as observed.

For starters, this looked like a triumph of the quark model — but soon enough, it wound up presenting a serious problem: quarks are spin 1/2 fermions, and therefore should, like all spin 1/2 fermions, be subject to a rule that Wolfgang Pauli discovered in the early days of quantum mechanics; it became known as the Pauli principle. It says that for a composite system in quantum physics, its quantum mechanical state must be antisymmetric under the exchange of two constituents.

Let us illustrate this principle by an example: If we look at a state that has the constituents A and B, calling it AB, we will obtain the state BA if we interchange these two constituents, another state. If A and B were objects with spin 1/2, the Pauli principle says that there is no natural occurrence of either of these states — only the state $AB - BA$. This state changes its sign when we interchange A and B — it is antisymmetric, the state $(AB + BA)$ would be symmetric. The Pauli principle does not permit it, and it does not occur in Nature. This Pauli "veto" is a subtlety of quantum physics, and it is of considerable importance for atomic physics.

But does it also apply to the quarks? Recall that we noticed the Δ^{++} baryon has the quark structure uuu. If we exchanged two of these quarks, nothing happens — meaning the state is symmetric, and should therefore not occur in our world. This means there must be a conflict between the simple quark model and experiment. And indeed, this difficulty was one of the reasons why the quark model was initially not accepted by many physicists. It was not before the early 1970s that the problem was solved. It turned out that the quarks have in addition to their electric charge and their spin, an additional property — a new index or charge. Take the u quark: it has electric charge 2/3 and spin 1/2; this new charge implies that the u quark can turn up in three different varieties, which we might tag as a, b and c (or 1, 2 and 3). Now recall that the number 3 has special meaning when it comes to colors — because all colors can be composed on the basis of just red, green and blue. So, this new index was given the "name" of color index, the new quantum index or color quantum number. To be sure, this has nothing to do with actual colors — it is just an allegorical naming of a new quantum number which appears to have great significance for quarks.

On the basis of this new degree of freedom, we have new ways of constructing a physical system from three quarks. Let us say, we write $(u_r u_r u_r)$, designating a system that contains three red up quarks. This gives us another conflict with the Pauli principle: this state is, again, symmetric under the exchange of any two quarks. There is one and only one state that is antisymmetric under any such exchange, and that is the state $(u_r u_g u_b - u_g u_r u_b + \cdots)$. By interchanging two quarks at a time and also changing the sign (from $+$ to $-$, or vice versa), we can come up with a total of 6 combinations, and the resulting states will be antisymmetric under the exchange of any two quarks. And we see that such a state has the property of containing all three colors, putting none of them at a disadvantage.

The introduction of this color degree of freedom permits us to take another bow toward the Pauli principle: just as in quantum electrodynamics and in atomic physics, this principle describes whether a possible quantum state is admissible or not. Of course, anybody can start considering a state consisting of three "red" quarks, as we did above, but the Pauli principle has the last word: no, not possible. But when all three quark colors are represented, its decision is: yes, admissible!

Suppose we interchange the colors r, g and b by the prescription, say, r → g, g → b, b → r, quite evidently, that will change nothing about the state we start out from. It is, as we say, invariant under this (or an equivalent) transformation. The rotation of the color degree of freedom defines another symmetry which is quite similar to the phase rotation in QED. In the year 1972, Murray Gell-Mann and this author proposed the introduction of color symmetry in analogy to phase symmetry in quantum electrodynamics. This simply defines another gauge theory, following the example of QED, and consequently called quantum chromodynamics or QCD (using the Greek word for color, "chromos"). Originally regarded with a fair amount of skepticism, QCD turned out to be the theory to give us excellent new access to the dynamics of the quarks inside nucleons.

But before going into details, let me point out an important difference between QED and QCD that is due to the symmetries they are based on. The gauge symmetry of QED is actually a very simple one: it is the symmetry of phase rotations. The visible phenomena do not care whether we rotate the system by 30°, turn it back by 10°, then forward again by 37°, and so on. Any rotation is easily described by one parameter: the angle of rotation. If we perform two such rotations in series — say, first by 10°, then by 30° — we will wind up with the system rotated by 40°. If we change the order of these rotations: first by 30°, then by 10°, we again wind up with 40° total. The result does not depend on the particular sequence. The mathematical term for this kind of symmetry is called an Abelian symmetry. It was named after the 19th century Norwegian mathematician Niels Henrik Abel who was honored by having his name changed into a new adjective in mathematical lingo.

Color symmetry is not so simple, and we can visualize this by an example from geometry. Rotations in a given plane in, say, a two-dimensional space, are an example of an Abelian symmetry. Once we introduce a further dimension so as to come up with three-dimensional space, we can enlarge the symmetry so as to include rotations in this space. Now we have three different rotations that are independent of each other — the rotations about the x-axis, about the y-axis and about the z-axis. That is why an arbitrary rotation now needs three parameters for its definition. There can be any combination of rotations. We can start with a 10° rotation about the x-axis; to be followed by a 20° rotation about the new y-axis; or we perform the 20° rotation about the x-axis and only then a 10° one about the new y-axis. In both cases we wind up with a rotation of three-dimensional space. But it turns out — as we can easily check by doing a little experiment or an easy calculation — that the results of our different sequences of rotation are not identical: a rotation about the z-axis tells them apart.

What it comes down to is the sequence of the individual rotations. This kind of symmetry is called a non-Abelian symmetry. And a gauge theory that results from a non-Abelian symmetry is, in consequence, called a non-Abelian gauge theory. Theories of this kind were first looked at in the 1950s by the US theorists C. N. Yang and R. Mills. But the first notions for building up such theories were contributed some ten years earlier by Herman Weyl in the United States, Oskar Klein in Sweden, and Wolfgang Pauli in Switerland.

In QCD, we are dealing with a non-Abelian gauge theory. But the rotations in color space are more complicated in simple geometric space. Given that both quarks and electrons are described in QED by complex fields, color symmetry results when we replace x, y and z, by complex axes. The mathematical idiom for such a theory is the group-theoretical term SU(3) symmetry. This may sound worse than it actually is — but the number of different transformations needed to accomplish arbitrary transformations in the quarks' color space is larger again: it is eight. That implies we need eight parameters to pin down each and every transformation color space, five more than in geometric three-dimensional space. We would be correct if we designated QED as a theory that has one color only. In fact, of course, we do not speak of color in QED, we speak of electric charge.

A layman will have no trouble understanding why the number eight comes up high in QCD. We have three colors, and transformations can be characterized by mentioning color → color. They are r → r, r → g, r → b, g → r, g → g, g → b, b → r, b → g, b → b, a total of nine possibilities. Note that we expressed include the cases where colors do not change. That is just like in QED, where that one is the only possibility. But the one case, where all colors remain the same, should not be counted — so that we are left with eight possibilities. If there were two instead of three colors, we would have a total of $4 - 1$, or 3, just like for normal rotations in space.

If we now admit color symmetry as a gauge symmetry and postulate that the related field equations respect it, we wind up, as in QED, with the interaction of the quarks with quanta that take the part of the photons in QED. We are now about to find out that those field quanta are responsible for the strong forces of quark interaction, i.e., for the binding of quarks inside the nucleons. We might say they provide the glue that holds these together — and so they were given the amusing name gluons. This is analogous to the photons that carry the electromagnetic force which acts on electric charge. In a similar way, the gluons provide the force that acts on the color charge of the quarks. The specific nature of the color gauge theory, this interaction differs from what happens in QED. The first experimental indications of the gluons' existence were found at the German research center DESY in Hamburg in the late 1970s.

Let us stress that a non-Abelian gauge theory differs in its structure from that of an Abelian one, such as QED. In particular, the coupling of photons to electrons differs from that of gluons to quarks. When an electron interacts with a photon, the particular state of the electron, particularly its momentum, but it remains an electron. By comparison, the interaction of a quark with a gluon is able to change the color of the quark. A red quark may well change into a green one, etc. These interactions are given when we consider transformations in color space, as we mentioned above. And as there are eight different color transformations, we need a total of eight different gluons. These quanta can then be characterized by the manner in which they "transport" color charges. So, there is the red $\rightarrow$ green gluon, the green $\rightarrow$ blue gluon, and so on.

In this vein, let us use the analogies of QED and QCD to gather the relevant features of the two theories:

QED	QCD
electron, muon	quarks
electric charge	color charge
photon	gluon
atom	nucleon

As we stressed above, one important difference between QED and QCD is the number of different quanta that transport forces: just the photon in QED, but eight gluons in QCD. The electromagnetic interaction does not change the electric charge of the particles involved; but the color of a quark can very well change due to one gluon interacting. There is another important difference: photons are electrically neutral, i.e., they cannot have an electromagnetic interaction with their likes. This is an important property for both Nature and for technology. Myriads of photons traverse space as part of a laser beam, traveling together at the speed of light without any mutual interaction. This is not possible for gluons; it was soon noticed that they interact not only with quarks, but also with each other. Just like the quarks, the gluons carry a color charge. That's easily noticeable from the fact that it is possible to tell them apart due to their "color". A gluon which turns a red quark into a blue quark is clearly a different creature when compared with another one that changes a red quark into a green one.

The fact that gluons themselves have a color charge has dramatic consequences for the dynamics of quark and gluon interactions. In particular, it influences vacuum polarization.

The vacuum is, according to the theoretical framework of QCD, replete with virtual quarks, antiquarks, and gluons. Looking at the surroundings of a quark, we notice that they are changed by the chromodynamic interaction. Just like in electrodynamics, there is vacuum polarization. Because of its color charge, the quark chases the virtual quarks from its surroundings but attracts the antiquarks. This effect is somewhat comparable to what happens in electrodynamics: the effective color charge of the quark is partially screened off. But since the gluons also carry color charge, the ocean of virtual gluons around the quark is also being polarized. This effect does not exist in QED. An electron is surrounded by virtual photons, but those are not influenced by the electron charges — simply because they do not carry charge themselves.

When, in the early 1970s, the effects of vacuum polarization in non-Abelian gauge theories were investigated, it was assumed that there was a similarity to QED, so there would be a charge screening. There was a real surprise when it was found that the effect is qualitatively different. First calculations were done by Russian theorists and by their Dutch colleague G. 't Hooft. Centering their interest on QCD, D. Gross, Frank Wilczek and David Politzer evaluated effects of vacuum polarization. They found out that virtual gluons tend to stick to quarks. That way, they generate a strengthening of the color charge at longer distances. This leads to an immediate strengthening

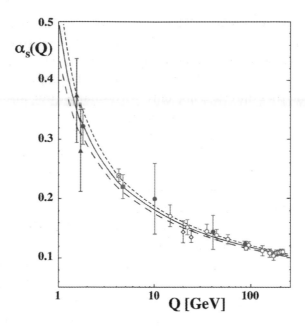

Fig. 7.1. Behavior of the chromodynamic coupling constant as a function of energy.

of the interaction. In analogy to the QED "fine structure constant" α, the strength of the color interaction is designated as α_s (or a strong).

We already mentioned that the electromagnetic fine structure constant increases at very small distances (so it is not really a constant!). In QCD, the opposite is the case: due to the gluons' contribution to vacuum polarization, the QCD "constant" α_s decreases at small distances (see Fig. 7.1). So, again, we have a contradiction in this term, α_s is not a constant, it depends on the distance and the energy of the process in question. Still, we tend to speak of the "constant of interaction", because α_s always has the same values at a given distance, irrespective of the specific experimental context. Indeed, α_s was determined in many experiments in electron–proton interaction or in electron–positron annihilation. We might mention the value of α_s at an energy of about 90 GeV, which is about $\alpha_s = 0.12$. That means, the QCD interaction is slightly stronger at this energy than the QED one. But still, α_s is small when compared with 1 so that we succeed, in the framework of a perturbative calculation, to calculate processes of quark–gluon interactions.

But α_s increases at low energies: the interaction increases, indeed, with decreasing energy but increasing distance, so that perturbative methods can no longer be applied. That is because the gluon cloud around the quark increases continuously, implying that the energy density of virtual particles becomes more important. When summed over all space, it would add up to the mass of a quark. But calculations show that the mass of a quark then increases to infinity, so that talking of a "mass" no longer makes sense. This, again, implies that, in QCD, quarks cannot be seen as free physical quantities, they are permanently bound inside the nucleons. Still, it makes sense to speak of quark masses: they are parameters which depend on distances of observation, just like α_s.

It is instructive to try and imagine the interaction of α quark and its antiquark. At small distances, that is, at distances that are small when compared with the characteristic scale of the strong interaction $(-10^{-13}$ cm), the force resembles in its range the electric one, diminishing with the square of the distance. Chromodynamic forces, just like the electric ones, can be depicted by field lines. But as the distance becomes larger than 10^{13} cm, things change: since gluons interact with each other, there is a force that acts from field line to field line to the effect that they attract one another.

This effect is comparable to the case in electrodynamics, where two parallel wires that carry an electric current which moves in the same direction attract each other. The relevant force is due to the magnetic field surrounding the conducting wires. In QCD, the forces between the gluonic field lines are due to the chromodynamic field lines that delineate the interactions between the gluons.

When we move quarks further apart, the resulting field lines look very different from that in electrodynamics. The lines that issue from one of the two quarks still connect to the other one, almost in parallel, resulting in something like a hose full of field lines — somewhat similar to the electric field lines between the two planes of a condenser. That similarity extends to the forces acting between two quarks and those between two condenser plates: it is constant, no matter what the distance. Whereas the electric force acting between a quark and an antiquark decreases with increasing distance, the chromodynamic force remains constant. This makes it obvious that we cannot separate the quarks from each other.

On the other hand, the forces between quarks become quite weak when the quarks are closely neighboring in location. This fact explains the observation that quarks behave like point charges when scattering electrons — the electron penetrates the proton deeply, since it is not at all influenced by gluonic QCD forces. It will interact with only one quark, and the interaction time is so brief that a gluonic interaction will be hardly noticeable.

It makes sense to compare that behavior with an everyday experience of ours: suppose we observe three colored spheres: one red, one green, one blue, that are bouncing around rapidly inside a hollow glass sphere, being scattered back off the walls. On the average, the time elapsing between two such scatters is only $1/100$ second. Let us assume we take a picture of the sphere and choose a relatively long exposure time of, say, $1/5$ second, we will not see the sphere. Instead, we see a blurred image when the colors of the spheres overlap, so that, on the average, the color appears to be white. If we choose a short exposure time — say, of $1/1000$ second — we will gain a clear image of the spheres and their colors.

In particle physics, quarks take the clue from those spheres: in high-energy scattering experiments of electrons, we can distinguish the three quarks quite clearly — simply because the exposure time, as defined by the collision time, is very short indeed. If we use, on the contrary, low-energy electrons, making the collision time a lot longer, nothing more than, at best, a blurred image will result. That implies the quarks act like free particles in short exposure, like deeply bound objects in long exposures. The physicists' term for the strong decrease of inter-quark gluonic forces is cases of tiny spatial or temporal distances as "asymptotic freedom". As a consequence, the interaction "constant" α_s decreases at smaller distances or increasing energies — so that it becomes possible to apply calculational methods of perturbation theory just like in QED. Note, however, that this statement applies only for processes that happen between quarks and gluons — not when strong binding between the quarks is prevalent. For the discovery of asymptotic freedom the Nobel Prize 2004 went to Gross, Politzer and Wilczek.

It is very interesting to look at the contribution of QCD to our understanding of strong nuclear forces inside atomic nuclei. The nuclei consist of nucleons, the nucleons of quarks. Particle physics was developed originally because physicists wanted to gain an understanding of nuclear forces. Why, for instance, are there forces that make six protons and six neutrons join up to form a remarkably stable nucleus, that of carbon? Today, we know that these are not at all fundamental forces; they are indirect consequences of gluonic interactions inside the nucleons — comparable to the forces between electrically neutral atoms that join up to build molecules. Those are indirect products of electric forces inside the atoms.

Before we quit our discussions of QCD, let us review what this field has to tell us about the observed structure of the particles we observe in Nature — particles that consist of quarks. Neither the quarks nor the gluons that bind them together exist as free particles because of their color degree of freedom. Given that there are three colors, the quarks are color triplets. And since the three colors are of equal importance, and interchangeable, we talk about color symmetry; mathematically, or rather in group theory, we describe this symmetry by the group SU(3).

In the above, we discussed that the colored quarks cannot exist as free particles. This is true in QCD not only for quarks, but for all objects that are colored, such as the eight gluons which form a color octet. The only particles that can exist as free ones are not colored. Mathematically, they are color singlets; their constituents are such that their colors cancel each other. This is an effect familiar from QED: atoms are objects that consist of electrically charged building blocks, the nucleons and the electrons in their shell. Altogether they are electrically neutral, and we are correct in calling them charge singlets.

The simplest color singlets that can be built up of colored quarks are structures that we have not considered yet. A quark together with an antiquark can build up a color singlet simply because they cancel each other's color. These objects that consist of matter and antimatter in equal parts do exist in Nature as unstable particles; they are produced in particle interactions, and all of them decay shortly afterwards. We call them mesons. The first observation of a meson was made in 1947, in a cosmic ray investigation. These particles were electrically charged but had no spin; their mass turned out to be 207 times that of the electron, about 140 MeV. These particles, that were given the name mesons ("the middle ones") are considerably less massive than the protons — so as a mass scale, they are indeed in the middle, justifying their name. In the same vein, nucleons and all objects consisting of three quarks are called baryons ("the heavy ones" — from the Greek word for "heavy").

It is, of course, possible to build up other color-neutral particles, or color singlets. The state we mentioned above, Δ^{++}, consists of three u quarks with the structure $(u_r u_g u_b - u_g u_r u_b + \cdots)$ fulfilled the antisymmetry requirement of the Pauli principle. We see that it also satisfies the property of giving all three colors equal weight. The state is seen to be antisymmetric when two colors are exchanged. Mathematically speaking, this makes it a color singlet. So, our color trinity permits us to construct a color neutral object from three quarks. In electrodynamics, which, by comparison, we might interpret as a gauge theory with only one "color" degree of freedom, electric charge, this would not work. Electrically neutral states, such as atoms, can be put together in electrodynamics only if we compensate every positive charge (of a proton) by a negative one (of an electron in the shell). This comparison illustrates that QCD is much more richly structured; it has many interesting properties that we will investigate in the following.

Chapter 8

Mesons, Baryons and Quarks

In the late 1950s, the physics of elementary particles resembled a zoo that was crowded with dozens of new particles that had been discovered in particle collisions. But then, at some point, physicists managed to change that chaos into order. They found out that there was a relatively simple scheme that permitted a description of the makeup of nuclear particles including the newly discovered ones: just imagine they are built up of three quarks. We have already seen that we need just two kinds of quarks, the u-quark and the d-quark: $p = (uud)$. This composite is color singlet, so its full description is $p'(u_r u_g d_b + \cdots)$. Chameleon-like, the proton jumps from one color combination to the next, but on the average, the colors cancel each other; as a result, the proton is a color-neutral state from outside.

As far as its spin is concerned, it has spin $1/2$, just like each of its quarks. But we know that spin is a directional quality; so, the proton spin is built up of two quark spins that have opposite direction so as to cancel each other, whereas the third quark spin determines the proton spin.

In the neutron, the parts played by the u-quark and d-quarks is interchanged. This is due to a higher symmetry which was discovered by Werner Heisenberg in the 1930s, shortly after the discovery of the neutron; he called this new degree of freedom "isospin", mathematically analogous to spin. This is the first instance of an internal symmetry: it is marked by the fact that, for a symmetry operation, external, i.e., geometric — properties do not run the show; only internal properties are important, the substructure of the nucleons. In the framework of QCD, isospin symmetry emerges quite naturally: any two quarks include their color degree of freedom in their strong interaction — their interaction does not differ. The only difference of u-quark and d-quarks is due to their electromagnetic interaction and to their masses. They form what we call an isospin symmetry in Nature.

Also in the phenomenology of the π mesons we mentioned preciously, isospin symmetry adds its mark: the electrically charged pions have a very simple substructure. They consist either of $\bar{u}d$ (which make up an electric charge of -1) or $u\bar{d}$, for a positive charge. Quite simply, the positive and the negative pions are each other's antiparticles. Isospin, however, postulates the existence of our electrically neutral pion with roughly the same mass, such that, altogether, there exists a triplet consisting of π^+, π^0, π^-. And indeed, the neutral pions were discovered soon after the charged ones. And their mass is very close to that of their charged partners, about 140 MeV.

It turns out that the substructure of the π^0 has some added complexity. We might think we should be able to build up two neutral mesons, based on $\bar{u}u$ and $\bar{d}d$. Now, which is the one that has long been seen? The obvious answer is: some linear combination of these two. Quantum theory comes to the rescue. The π^0 is one-half $\bar{u}u$, one-half $\bar{d}d$. And by implication the neutral π meson is its own antiparticle, just like the photon. Consisting in equal measure of matter and antimatter; it is very unstable. It is seen to decay right after having been created in some particle reaction, almost always into two photons. The decay is an electromagnetic one — the quarks and their antiquarks radiate off all their energy into photons.

The charged pions have a much longer lifetime, about one hundred-millionth of a second — which is relatively long in the particle physics context. It may be instructive to compare the distance they traverse in their lifetime with the distance light travels in the same amount of time. For a charged π meson, that distance is just 7.8 meters. But a charged pion may also travel across about 100 meter before decaying (assuming it is a high-energy one). At the bottom of this mystery is Einstein's relativity of time. The "light path" of the neutral pion is orders of magnitude smaller, and measures just about 100 atomic diameters.

If the charged meson has a lifetime much longer than that of its neutral cousin, that is due to the fact that its decay occurs by means of the so-called "weak interaction" — similar to the process which mediates the radioactive decay of many unstable atomic nuclei. We will notice later on that the electromagnetic and the weak interaction are closely related. The charged π mesons decay into two particles — a charged muon (the "heavy" brother of the electron) and a neutrino, a very light, electrically neutral partner.

It is not as though nucleons and π mesons were the only particles we can build up from quarks in the framework of QCD. We already mentioned the doubly charged Δ^{++} particle, often called the delta resonance, as an example. It is built up of three (uuu) quarks. Its spin of 3/2 units is a simple addition of three aligned quark spins. By systematically replacing up quarks by down quarks, we see that there must be four charge states with the substructures (uuu), (uud), (udd) and (ddd). The electric charges must then must be $(+2, +1, 0, -1)$. They form an isospin quartet.

The Δ particles are, we might say, excited states of the proton, and we can produce them by running a pion beam into a nucleon target. They have short lifetimes, similar to the time in which light traverses a nucleus; their decay products are mesons and nucleons. Their lifetimes are so short, we might say, that calling them particles is a misnomer, since a "particle" is defined largely by its mass. But we know that quantum theory establishes the uncertainty relation between mass and lifetime. The shorter the time between creation and decay of a particle, the poorer the definition of its mass. To be more precise, we cannot measure a mass properly, but just a median mass value — averaging over many production instances of particles with the same composition and quantum numbers. This median mass for the Δ baryon is 1232 MeV, almost 4/3 of the proton mass. The decay width of the Δ, as we call its mass uncertainty, is almost 120 MeV, about 10% of its mass.

Now, we know that the Δ and the proton have the same quark substructure (*uud*), so we might ask: what's the most important difference? It turns out this is a matter of their spins only. In the Δ case, the three spins are aligned, but not so in the proton case. To convert a proton into a Δ, we would have to turn around one spin orientation. But that operation needs energy input — exactly as much as what corresponds to the mass difference of almost 300 MeV.

In addition to the Δ particles, we have quite a few unstable particles consisting of three quarks: they are objects the quarks of which have orbital angular momentum in addition to their spin, or that are in excited quantum states in analogy to what we see in atomic physics. The same is true for mesons: π mesons do not have angular momentum, because the spin of quark and antiquark compensate each other. If we turn around one quark spin, we obtain a particle of spin 1 with a considerably higher mass: it is the so-called ϱ (Greek letter rho) meson with mass 770 MeV. Just like the Δ baryons, the ϱ are very unstable and decay immediately after their production in a particle collision, into two π mesons. The width of the p meson is enormous: about 150 MeV, or some 20% of its mass.

Mesons also have many "resonances", where the individual quarks are in excited states, e.g., where they have relative angular momenta. Hundreds of such short-lived particles are mentioned in the "Particle Data Book" that is issued annually, including all new data. There is no space here for a more detailed description.

Just for the buildup of nuclear matter, we need only *u*-quark and *d*-quarks. To understand the mass values of various particles we deal with, we have to assign each quark type a mass of its own, be it quite small: It is of order 5 MeV, with the *d*-quark slightly heavier than the *u*-quark. This latter difference is at the basis of the observation that the neutron is heavier than the proton.

In the last four decades of the twentieth century, particle physicists discovered many new particles that show evidence of four additional quarks. They are called *s* (for "strange", with charge $-1/3$), *c* (for "charmed", with charge $2/3$), *b* (for "bottom", with charge $-1/3$) and *t* (for "top", again with charge $2/3$). These former "flavors" of quarks, which we categorize as pair, analogously to

(u, d), as (c, s) and (t, b), can be building blocks of heavier matter. All of their masses are much larger than the small masses of u and d; they reach from 150 MeV for the s quark to 175,000 MeV for the t-quark. This means the single t-quark has a mass as heavy as an entire atom of gold which consists of 197 nucleons in most cases.

The new theoretical ideas about quarks took hold shortly after the memorable date of November 11, 1974, which is often called the "November Revolution" among particle physicists. There was an experimental breakthrough at the Stanford Linear Accelerator Center (SLAC) and at the Brookhaven National Laboratory (BNL) on Long Island: it was the discovery of a very unusual particle with a lifetime about 10,000 times longer than expected under the circumstances. This particle is a meson consisting of a c-quark and its antiquark.

Particles containing an s-quark (or "strange" quark) had been discovered as early as the 1950s, and they were called "strange particles", a long time before the quark model existed. If we take a neutron and replace one d-quark of this (udd) composite by an s-quark, we come up with the Λ (Greek capital "Lambda") particle, which is about 150 MeV heavier and remarkably, its lifetime is relatively long when measured by its "light path" of almost 8 cm. It decays into a nucleon and a π meson, in the process changing the s-quark into a u-quark. This exchange is a "weak interaction" process the likes of which we will describe in the next chapter.

In addition to the Λ particle, there are quite a few particles containing s quarks, they may contain two s-quarks (like uss) with no electric charge, or even sss, with charge -1. This particle, the Ω (capital Greek Omega), has a mass of about 1670 MeV and is somewhat analogous to the Ω resonance we discussed above. But its lifetime is much longer; its "light path" measures some 2.5 cm. This is due to the fact that the Ω can decay only if its s-quarks change into u-quark or d-quarks in the process. That, however, is possible only as a "weak interaction" process and not a strong nuclear interaction as in the decay of Ω particles. This fact explains the long lifetime of the Ω. Its 1964 discovery at Brookhaven National Lab set a milestone in the twentieth century particle physics, because its existence had been predicted, including its mass, in the framework of symmetry models for elementary particle properties.

We have also known mesons that contain s-quarks, for a long time, in particular the lightest one of them, the K meson, with a mass of 496 MeV. There are four different versions of it, according to the quark contents: $(\bar{u}s)$ are the K^+, $(\bar{s}u)$ the K^-, (ds) the K^0, $(\bar{s}d)$ the K^0. Just like the π mesons, the kaons are unstable: they decay in a weak interaction, changing the s-quark to a u-quark.

Isospin symmetry can be seen as a symmetry which results if we take two quarks — say, the u-quarks and d-quarks — as coordinates of a two-dimensional coordinate system. The QCD interaction does not depend on the specific quark type, and the masses of the two quarks we are dealing with do not make much of a difference; as a result, we can perform a rotation in this coordinate system without affecting the strong interaction physics of the system. This is synonymous with isospin symmetry. If the s-quarks were massless or nearly massless, we could even deal with rotations in a three-dimensional quark space spanned by u, d and s. True, the mass of the s-quark exceeds the u and d masses by about 150 MeV, so this symmetry is broken; still, this broken symmetry, called SU(3), which was first studied by Murray Gell-Mann and Yuval Neeman, serves well for the description of the hadrons built up out of u-quarks, d-quarks and s-quarks. It predicts, for instance, that the two nucleons proton and neutron have six more partners with which they form an "octet". These added states contain the Δ particle we mentioned before. Similarly, the Δ resonances we discussed join the Ω states and others to form an "irreducible representation" containing 10 members, which we call a decuplet.

While, in the early 1970s, it was believed that there is some deep significance to the observed symmetries, we are now convinced that these symmetries are simply a consequence of the quark substructure of the hadron spectrum. The resulting SU(3) symmetry functions fairly well simply because the s-quarks are only about 150 MeV heavier than the u-quarks and the d-quarks. If the mass difference were ten times that, we could essentially abandon the symmetry. This is easily illustrated if we substitute a c-quark for the s-quark in the Λ particle. The resulting particle with the substructure (*udc*) has a much higher mass value than the Λ because of the relatively heavy mass of the *c*-quark; it "weighs in" at 2285 MeV. When we consider a symmetry which contains, in addition to light quarks *u*-quarks, *d*-quarks and *s*-quarks, also the *c*-quark (which is our SU(3) symmetry), this heavy particle will appear, together with the nucleons *n* and *p*, in a particle family containing a total of eight particles, which we called an octet above. The mass differences inside this octet are of order 100% — which illustrates the fact that we are dealing with a strongly broken symmetry.

If we extend these concepts to particles that also contain bottom or *b*-quarks, we encounter even more strongly broken symmetries. Actually, the mass differences within one grouping or family are now so large that the concept of a symmetry does not make sense any more. The mass of a baryon built up of the quarks *u*, *d* and *b*, winds up at about 5,600 MeV, i.e., at more than five proton masses.

A special role among the quarks is reserved to the top quark, for short *t*. Its gigantic mass of about 175 GeV (or 175,000 MeV) presented a real riddle to physicists. It is so heavy that it decays immediately after being "created" in an interaction; and with this vast mass, it has a huge amount of possible decay channels. That makes it decay even more rapidly. A bound hadron of quark content (*u*, *d*, *t*) cannot even form by the time of the decay. This is true although the time needed for the formation of such a hadron could be happening within the minuscule time that it takes a Δ particle to decay; its decay is still faster. As a result, that quark is only one of the six quark flavors that cannot act as a building block of hadronic matter. It remains just a phantom in high-energy particle collisions, just decaying immediately after creation — leaving a *b*-quark and many other particles in its wake.

Today, as we have seen the veils coming off the secrets of atomic nuclei and nuclear constituents, we observe a picture of nuclear forces and building blocks that, on one hand, is comprehensible in the framework of quantum field theory, but that is of an obfuscating complexity. True enough, our Nature needs only two quarks, *u* and *d*, to build up stable atomic nuclei; but the prevailing symmetry structure implies evidence for six quarks. Perceived from the outside, a stable atomic nucleus looks like a solid citizen of Nature; but way inside there is a boiling microcosm — a world of complex, unstable hadron interactions becomes visible if not fully intelligible, manifesting its existence in every high-energy particle collision.

Chapter 9

Electroweak Interactions

Joining the electromagnetic force that acts between leptons and quarks there is another important interaction between the same participants — the so-called weak interaction. This is the phenomenon responsible for the fact that a neutron is not stable, but is subject to "beta decay" into a proton, an electron, and a neutrino:

$$n \rightarrow p + e^- + \nu_e \, .$$

All reactions that are due to the weak interaction can be attributed to one of the two categories: either to those where the electric charge changes — such as in "beta decay", where a neutral neutron changes into a charged proton (as we just saw above), or to those where there is no change in electric charge, as in neutrino scattering

$$\nu + p \rightarrow \nu + p$$

The first category is called a charged-current weak interaction, and the second a neutral-current one. One thing they have in common is the participation of four particles, four fermions. Sometimes one fermion changes into three fermions, as in neutron beta decay, and sometimes two fermions interact in a process where they may form a different fermion pair, as in

$$\nu_\mu + n \rightarrow \mu^- + p$$

An identifying feature of all weak interaction processes is that they are characterized by a constant that indicates their strength. This parameter, called the Fermi constant after the Italian physicist Enrico Fermi, has been experimentally determined with great precision. It is a small quantity of about 1.16×10^{-5} GeV^{-2}. Note that it is not a dimensionless parameter such as the "fine structure constant"; rather, it is of dimension inverse energy-squared. Its small numeric value is responsible for the "weakness" of the weak nuclear interaction where we compare its strength with electromagnetism.

Still, theoretical physicists do not like to deal with a dimensional quantity. The implication is that we cannot come up with a consistent theory which is able to describe quantized phenomena. The reason is that it carries a characteristic energy scale of 1.16×10^{-5} GeV^{-2} or $(0.294$ GeV$)^{-2}$. This means the "weak interaction" is marked by an energy of 294 GeV. And precision studies of this phenomenon tell us that our picture of the weak interaction must break down at energies large in comparison with this energy scale — that something new must come up.

The simplest new feature would be to imagine the weak interaction between four particles to be similar to the electromagnetic interactions between two electrons by the exchange of virtual particles. Let us take the "weak" decay of the neutron and suppose that the proton can be changed into a neutron by emitting a virtual particle which we will call a W^-. It will be very short-lived, being just a virtual entity and will then decay into an electron and an antineutrino.

We can see this process to be quite analogous to the electromagnetic interaction between two electrons, which happens by way of the exchange of a virtual photon. Just replace the virtual photon by a W-boson — but recall that the W is electrically charged, whereas the photon is electrically neutral.

We can describe the processes involving neutral currents quite analogously — we just need a new neutral particle that will be exchanged among the relative fermions; let us call this particle the Z-boson. If, for instance, matter scatters a neutrino, the neutrino remains after having exchanged a Z-boson in a weak interaction process.

Note that the description of weak interaction processes in terms of the exchange of virtual particles simplifies matters considerably: the elementary interaction is no longer the process of an interaction among four fermions, as formulated by Fermi; rather, it is now the interaction between two fermions and the virtual boson. That puts the weak interaction into a formal analogy with electromagnetism. Now we can characterize this interaction by a simple dimensionless number.

But how can we discuss the Fermi constant in the framework of this theory? It is easy to see that the W-boson and the Z-boson have to be massive. The mass of the W-boson is closely related to the Fermi constant: the critical energy of 294 GeV which we mentioned above is simply the mass of the W-boson divided by the constant of interaction between fermions and W-boson.

But before we turn to the masses, we have to consider the strength of the interaction proper. Given that the mediation of the weak interaction is carried by the weak bosons, quite analogous to QCD, we have to answer the question: is the analogy between weak interaction and electromagnetic interaction just a formality, or is there a direct connection between them? Such a connection could exist if, for instance, the weak interaction were not really weak, but showed up as such only at relatively low energies simply because of the large masses of W-bosons and Z-bosons. If, on the other hand, we assume that the elementary interaction of fermions and W-bosons is equal to the electromagnetic interaction of electrons and photons, we simply wind up with a problem. In this case, the mass of the W-boson has to be 37 GeV — a value, which has long been excluded as being way too light.

At this juncture, we have to mention another important difference between electromagnetic and weak interactions as far as space reflection is concerned: Take an arbitrary fermion — say, an electron or a quark. We can construct this fermion from a right-handed and a left-handed fermion. Now, a right-handed fermion is a particle that has its spin pointed in the direction of its momentum — we might compare it with a screw that turns to the right — whereas a left-handed fermion has its spin pointing in the direction opposite to its momentum.

Now let us look at a reflection in a mirror: a left-handed fermion changes into a right-handed fermion. If Nature were invariant under space reflection, left-handed and right-handed fermions would have to show exactly the same interactions — and this in fact is the case for electromagnetic processes, but not for weak ones. As early as in 1956, experimental evidence was found that the weak interaction violates space reflection symmetry; this discovery entered into the annals of physics as "parity violation". It was first observed in the weak decay of cobalt into nickel.

It was soon found that parity violation by leptons and quarks is quite easily described as long as we limit ourselves to interactions of fermions with W-bosons: only left-handed leptons and quarks interact with W-boson; right-handed ones do not. That means the interaction of W-bosons with fermions is quite different from that of the fermions with photons. To this day, we have no notion why Nature chooses to be left-handed when interacting weakly. We have to accept this fact without understanding it.

We mentioned above that W-bosons can change an electron into a neutrino, and vice versa. This fact reminds us of a similar connection in chromodynamics: in QCD, gluons are capable of changing one "color" of quarks into another one — say, making a green quark out of a red one. This works because color transformations act like "charges" of the color group $SU(3)$ — the gauge group of strong interaction theory. Do we have an implication here that W-bosons are gauge bosons in analogy to the gluons? What gauge group would that point to?

This last question can be readily answered: as far as the weak interaction is concerned, leptons and quarks always act as doublets. The left-handed electron neutrinos ν_e and electrons, the up and down quarks, etc.

$$\begin{pmatrix} \nu_e \\ e^- \end{pmatrix}_L, \qquad \begin{pmatrix} u \\ d \end{pmatrix}_L \quad \text{etc.}$$

or simply all the doublets of left-handed fermions. The implication is that we have to consider all transformations concerning doublets — implying the group SU(2), in contrast to the group SU(3) in strong interactions. The latter concerns three charges. Two of the latter can be identified with the weak charges we have introduced — the charges that change the upper component of a doublet into a lower one, or vice versa — say, $u_L \to d_L$ and vice versa. These "charges" change the electric charge by one unit. In addition, there is a third "charge", an electrically neutral one. How should this be interpreted? Is it simply a "weak charge" we need for a formal description of a neutral current interaction, or are matters more complicated than that?

Before we address this question, let us deal with another problem: We aim at interpreting the W and Z bosons as the gauge bosons in a gauge theory of the weak interaction, which will be analogous to QCD. But: the gluons of QCD are massless like photons, whereas W's and Z's are quite massive as we have seen, and that difference presents a severe problem. It is impossible simply to enter the massive aspect of gauge bosons into our gauge theory without, as a consequence, coming up with abstruse results.

Fortunately, there is an interesting possibility to hold on to the massive quality of the gauge bosons without introducing it explicitly. It was discussed by a number of theorists as early as in 1965, most explicitly by Peter Higgs. The ruse that does the trick is the introduction of scalar bosons in addition to the vector bosons. They are postulated to interact with the gauge bosons, with the consequence, among other things — that this interaction "gives" the gauge bosons a mass. This process has an added implication: it violates the prevalent gauge symmetry — leading to the concept of "spontaneous symmetry breaking". M. Veltman and G. 't Hooft were able to show, in 1971, that the introduction of mass parameters by means of this mechanism does not lead to senseless results. Their brilliant interpretation was recognized by the Nobel Committee in 1999.

The simplest trick for the construction of a gauge theory of weak interactions implies the use of the gauge group SU(2). That makes the left-handed leptons and quarks doublets, the right-handed ones singlets that do not participate in the interaction. This scenario gives us three gauge bosons, which we take to be W^+, W^- and Z^0. All three of these have the specific property that they interact only with left-handed fermions.

In 1977, new evidence was collected on weak neutral interactions. It showed an interesting feature: This interaction, in contrast to the charged one, also interacts with the right-handed fermions. This new discovery excluded a simple SU(2) theory as we postulated above.

Now, we know that the electromagnetic interaction acts on the left-handed as well as on the right-handed fermions, similar to the neutral-current weak interaction. This observation suggests taking a look at a possibility to unify the electromagnetic and weak interactions in some way. A number of such attempts were made over quite a period of time — by Sheldon Glashow in 1962, by Abdus Salam and John Ward in 1964, by Steve Weinberg in 1967, and again by Salam in 1968. Glashow, Salam and Weinberg were honored for their work by the 1979 Nobel prize. Ever since, it has become clear that the framework of this theory does full justice to the weak interaction.

In a unified theory of electromagnetic and weak interactions — or, as we call them today, of electroweak interaction — we need to deal with a total of four gauge bosons: W^--boson, W^+-boson, Z^0-boson and the photon. For that reason, we have to extend the relevant gauge group: the simplest extension is the inclusion of an added U(1) group, leading to the gauge group SU(2) × U(1) — a gauge group consisting of the product of SU(2) and U(1).

Suppose we start from this group and generate the masses of the gauge bosons by means of spontaneous symmetry breaking. In the process, we find some notable implication. The two W-bosons obtain some mass, which initially, can be chosen arbitrarily. The two electrically neutral bosons have an interesting mass spectrum: one of them winds up slightly heavier than the W-boson, whereas the other one remains massless. So it is natural to identify the last one with the photon, whereas the heavy one is the Z^0-boson. That means we wind up with a theory that couples the weak and electromagnetic interactions directly, and establishes a close link between photon and Z-boson.

Given that electromagnetism and weak neutral current are closely connected, it appears natural that the Z-boson interacts with left-handed and right-handed fermions. To be sure, there was no way to predict the weak neutral interaction with any precision. It depends on a parameter that is not implied by the theory, but has to be measured by experiment. We have accustomed ourselves to characterize this parameter as an angle, called θ_w where the subscript w stands either for "weak" or for "Weinberg", after Steven Weinberg, whom we mentioned above. This angle tells us the strong interrelation between electromagnetic and weak interactions. For the special case with $\theta_w = 0$, there would be no connection — but experiment shows us that, in fact, $\theta_w = 28.7$.

Our $SU(2) \times U(1)$ theory of electroweak interactions makes an important prediction: the masses of W and Z bosons are fixed once we know the angle θ_w. Both of these bosons were discovered in the year 1984 at the European Center for Nuclear Research, (CERN), in proton–antiproton collisions. By this time, we have precise results and find $m(W) = 80.45$ GeV and $m(Z) = 91.19$ GeV, in close similarity to the theoretical predictions.

To study the Z-boson more precisely, an electron–positron collider, called LEP, was built at the CERN laboratory, starting operations in 1989. Over the next decade, some 20 million Z-decays were observed. In this way, the parameters of the elctroweak interaction were measured with high precision, giving us a close glimpse of the mass of the Z-boson and the Weinberg angle θ_w. Remarkably, the results obtained at LEP were quite close to the predictions of electroweak theory. Many theorists expected experimental results that would differ from their predictions, but that did not happen. Nature justified electroweak symmetry expectation. The LEP research program provided a splendid justification of the $SU(2) \times U(1)$ theory.

Finally, let us mention that new results of neutrino-initiated interactions are hinting at facts that have not been expected in the framework of the simplest $SU(2) \times U(1)$ models. There, we expect neutrinos to be massless. But several experimental indications, like those obtained with the neutrino detector that has been used for years near Kamioka in Japan, "neutrino oscillations" have been observed. This means that, say, a muon neutrino radiated off by the weak decay of a charged pion, can change its identity while flying through space: it may become a tau neutrino (ν_α). Such "neutrino flavor change" is possible only if neutrinos are not massless. In that case, there is no reason why a neutrino, say, a muon neutrino (ν_μ) is a fixed-mass state, with some given mass. It might be a mixture of two or even three mass states. Given that weak decays do not permit us to observe neutrino masses due to their very small size, there is no reason why this scenario should not be accurate.

On the other hand, the different mass states will be propagated at different velocities; this implies a change of neutrino structure as a function of mass states — such that a muon neutrino may well change into a tau neutrino, only to bounce back into the muon neutrino state, and so on. Such processes have been observed, if indirectly, not only near Kamioka in Japan, but also with the neutrino facility close to Sudbury in Canada. We must conclude that neutrino flavor mixtures exist in Nature, and that the relevant mixing angles may well be large. Neutrino masses, of course, must be quite small — on a scale we do not yet know with any assurance. But indications are that neutrino masses are below 1 eV.

Chapter 10

Grand Unification

It has become quite obvious that we have witnessed a major break-through for modern physics as the theories of electroweak interactions (or, SU(2) × U(1)) and of the strong interaction (or, SU(3)) were developed. Their framework provides ways to describe almost all elementary particle phenomena, covering weak, electromagnetic, and strong interactions. Still, there are more observed features that are not covered. One of these is the fact that the electric charges of leptons and quarks are fixed   or, as we say today, quantized. The electric charges of the electrons and muons are −1, those of the quarks 2/3 or −1/3. It appears as though a law we do not yet know forced the charges to assume these values, including the feature that the fractional charges apply to the quarks which are color triplets, whereas the electrons and muons — both color singlets — have integer charges.

It is easy to see that the SU(2) × U(1) theory will admit arbitrary charges due to its inclusion of a free parameter, the "Weinberg angle" θ_w. It would be easy to change the charges of u-quarks to $2/\pi$ instead of 2/3, a quantization of charges can be imposed only by fixing the Weinberg angle.

There is another problem: the strength of the "strong" interaction when we compare it with the electroweak phenomena. We might ask ourselves: Is there a theory which provides a correct description of the strong and electroweak interactions including their

relative strength? Notice that the chromodynamic analogue to the
fine structure constant α_s at the energy $E = m(Z)$ is about 0.12
(or almost exactly 1/8), whereas the electromagnetic fine structure
constant α at the same $E = m(Z)$ value amounts to 1/128, or a 16th
of the previous value. A unified theory of particle interactions would
have to explain that fact.

So, let us now try and build up a theory that describes the strong
and electroweak interactions jointly. The theory must contain the
product of three gauge groups, SU(3) × SU(2) × U(1). A unified
theory of all interactions can be built up if we embed these three
gauge groups into a larger group. Mathematically, this is not so
hard — but we have to take care that the fermions are properly
represented: the leptons have to appear as color singlets, the quarks
as color triplets. This necessity puts a severe squeeze on possible
groups that will satisfy this condition.

The smallest group that contains the color group SU(3) and the
group SU(2) × U(1) which also doing justice to the fermions, is the
group SU(5). Let us recall that the fermions of the first lepton/quark
family are

$$\begin{pmatrix} \nu_e \\ e^- \end{pmatrix} (e^+) \begin{pmatrix} uuu \\ ddd \end{pmatrix} (\bar{u}\bar{u}\bar{u})(\bar{d}\bar{d}\bar{d}) \,,$$

where we have explicitly mentioned the colors of the quarks. The
fermions of the other two families are similarly listed, so that we
wind up with a total of 15 fermions per family. Let us now group
the fermions into two systems

$$\begin{pmatrix} \nu_e & \vdots & \\ & & \bar{d}\bar{d}\bar{d} \\ e^- & \vdots & \end{pmatrix}, \quad \begin{pmatrix} uuu & \vdots & \\ & & \bar{u}\bar{u}\bar{u}, e^+ \\ ddd & \vdots & \end{pmatrix}$$

The first of these systems contains five fermions, the second ten.
Now, it turns out that these two fermion systems are two different representations of the group SU(5); this implies that every
SU(5) transformation has to follow mathematical prescriptions for
the transformations of its components.

These prescriptions also determine the electric charges: the group SU(5) has a total of $5^2 - 1 = 24$ charges (note that the larger a group, the more different charges it contains: the group U(1) has just one charge, SU(2) has three, SU(3) of the color space has eight charges).

Interestingly, the charges within a group add up to zero. E.g., the electric charges of the five fermions we are considering have to cancel each other once we add them up: Since the electric charge of the neutrino is zero, this gives us a relation of the electric charges of the electron and that of the d-quark.

$$Q(e^-) = \frac{1}{3}Q(d)$$

So, the group structure imposes precisely the electric charges we observe in Nature. And, of course, the factor 3 is the number of colors the quarks may have. Analogously, we observe the electric charge of the u-quark to be $2/3$ — because is has to be just one unit larger than that of the d-quark.

The SU(5) theory also permits to calculate the "weak angle" and the chromodynamic fine structure constant for which we find:

$$\theta_w = 37.8° , \qquad \alpha_s = 8/3\alpha \sim 1/51$$

These two values present a real problem: they do not at all agree with the values that were determined experimentally: θ_w is about 28.7° and α_s is not as small as indicated above.

There is another problem concerning the SU(5) theory: it has 24 gauge bosons, corresponding to the 24 charges. There are the eight gluons of QCD, the W^+, W^-, Z^0, and the photon, adding up to 12 gauge bosons. That leaves 12 more gauge bosons that must exist and cause new interactions hitherto unseen. Not only unseen, but they look odd; they could, e.g., transform a lepton into a quark. To be sure, that is possible in principle — given the SU(5) representations which contain quarks as well as leptons: elements of one and the same representation can always be transformed into each other by a group transformation.

One of the consequences of these new interactions implies that the proton is unstable and can decay into lighter particles — say, into a positron that takes over the positive electric charge, and a π^0 meson. Now, this kind of decay violates baryon number conservation. This decay occurs because the new interactions permit two quarks in the proton to transform into a positron and an antiquark.

In SU(5) theory, the lifetime of the proton depends on the masses of the new gauge bosons. The observed great stability of the proton implies that the mass of these particles must be enormous — at least 10^{15} GeV. Only if we accept that mass we assure a proton stability corresponding to experimental evidence indicating some 10^{31} years. Notice that this number is many orders of magnitude larger than the age of the universe which, today, is being estimated at 14 billion years (or 10^{10} years). The fact that we feel we can infer a proton lifetime much larger than the proven age of the universe is due to the observational method employed: proton decay has been searched for by looking at large amounts of protons, e.g., at many tons of water.

Given the large mass of such new gauge bosons, we are constrained to assume a possible unification of the strong, electromagnetic, and weak interactions at energies no lower than 10^{16} GeV, that is where SU(5) theory may turn effective. There is an interesting implication to the onset of a new energy scale (of 10^{16} GeV): should SU(5) theory be correct, the coupling strengths of the strong, electromagnetic, and weak interactions will all be the same at energies of more than 10^{16} GeV, since the different interactions are nothing but different manifestations of one and the same unified theory. We then have to expect that the weak mixing angle θ_w is almost precisely $38°$, and the relation $\alpha_s = 8/3\alpha$ characterizes the strengths of strong and electromagnetic interactions.

Still, we have to keep in mind that our knowledge of coupling strengths is based on experiments performed at relatively low energies, say, below 10 GeV. There is no justification to equating these to coupling strengths we expect at energies of some 10^{16} GeV. We do expect, in the framework of quantum field theory, that the coupling strengths change slowly as a function of the prevailing energy scale. The QCD coupling strength diminishes with increasing energy because of the features of asymptotic freedom, whereas that of $U(1)$ theory increases slowly.

Experiments at the LEP rings at CERN permitted a precise determination of the parameters of $SU(2) \times U(1)$ theory. The fairly precise knowledge of the relative coupling strengths thereby permits us an extrapolation to higher energies. We find that the three coupling strengths approach each other at about 10^{15} GeV; but their energy dependence does not intersect in one point, as theory would suggest.

The dependence of coupling constants as a function of energy does indicate that the idea of a unification of fundamental forces makes sense. Still, it appears that somewhere on the long path from energies accessible today to unification energy, something must be happening beyond the slow change in coupling parameters, as a result of prevailing interactions. For instance, it might well be that, at energies of about 1000 GeV, new symmetries, and thereby, new interactions, appear.

Today, a number of theorists are discussing the onset of what is called supersymmetry. Symmetries as we have known them — such as, say, isospin symmetry — group particles of one and the same spin: a symmetry transformation can, for instance, easily change a proton into a neutron, but not a spin $-1/2$ proton into a spin -0 meson. But in the framework of supersymmetry, there may be a possibility to change a fermion into a boson. There may, for instance, be a spin $1/2$ quark that will be changed into a particle with spin 0 which, of course, does not exist in the observed spectrum. So, we are implying that there be a new particle with a mass high enough so it has eluded observation. This hypothetical particle is assigned the name squark, short form for supersymmetric partner of the quark. In fact, the supersymmetric extension of our Standard Model adds a new boson to every "old" fermion, and a new fermion to every "old" boson. The supersymmetric partner of the spin 1 photon is the hypothetical spin $-1/2$ photino.

To this day, we do not know whether Nature realized supersymmetry or whether it is just the theorists' extravaganza. If it exists, there must be a critical energy scale where this new type of symmetry sets in. That could also be the energy defining the mass scale for supersymmetric particles. The prevailing assumption is that super-symmetry sets in, if at all, at about 1000 GeV.

Should supersymmetry exist, the coupling strengths will change at energies where it appears — due to the fact that the supersym-metric partners will participate in interaction, contributing to the changes in c-coupling parameters. It can be shown that the exis-tence of supersymmetric partners at about 1000 GeV sees to it that all coupling strengths approach each other at about 1.5×10^{16} GeV (see Fig. 10.1). That means, we can formulate a supersymmetric variant of SU(5) theory that is consistent with the experimental data observed to date. We also find in their connection that the super-symmetric version of SU(5) theory leaves the proton unstable, but a bit less so than the non-supersymmetric version — with a lifetime of about 10^{33} years. This relatively long lifetime is compatible with data obtainable now.

We should not see more in SU(5) theory than an example for a theory that actually unifies the three observed interactions. But

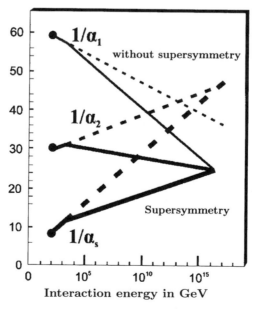

Fig. 10.1.   The behavior of coupling constants with and without supersymmetry.

it is not the only theory to do so. In particular, there is a theory based on the symmetry group SO(10) which describes the symmetry in ten-dimensional space. Compare this with SO(3) symmetry that, in three-dimensional space, describes all rotations. Now, SO(10) symmetry has the remarkable property that it has a representation containing 16 elements. This fact opens up the possibility to describe all fermions of a "family", including their antiparticles, in one and the same symmetry representation. Take, for instance, the fermions of the first family

$$\begin{pmatrix} \nu_e & : & uuu & : & uuu & : & \nu_e \\ e^- & : & ddd & : & \bar{d}\bar{d}\bar{d} & : & e^+ \end{pmatrix}$$

Clearly, the SO(10) theory is more comprehensive than the SU(5) version. Rather, it contains the latter as a partial symmetry, and it has 45 gauge bosons. There is also the interesting fact that we can attain a unification of coupling constants without having recourse to supersymmetry. This is due to the fact that SO(10) has more gauge bosons than SU(5) — including partners of the W-bosons that act on right-handed fermions. We have to add that these particles are considerably heavier than the customary W-bosons, so that they are not observable with presently existing means. At high energies, however, these particles are present to modify the coupling parameters. We arrive at a convergence of the coupling parameters if the masses of those new gauge bosons are of order 1000 GeV. This fact implies that SO(10) theory is as consistent with today's observables as the supersymmetric version of SU(5). The future will tell whether Nature actually has recourse to these theoretical possibilities. These theories share the feature that the proton is unstable, with a lifetime not much larger than the limit imposed by experimental evidence obtainable today.

Should the proton actually be able to decay into leptons and photons, the implication would be that baryon number is not exactly conserved. That would help to explain one of the oldest phenomena in our universe. Matter in our world consists mostly of nucleons, and thereby of quarks. Antimatter is made up of antiquarks, but is not observed in our stable galactic system. We also have hints that distant galaxies are made up of matter — not of antimatter. That tells us that the baryon number of the visible universe is enormously large. And we know that the universe such as we observe it originated some 14 billion years ago, in the Big Bang — when matter was created in an extremely hot phase. If baryon numbers were strictly conserved, it would have been the same at the end of the Big Bang as it is today. That means, our cosmos must have been born with a large baryon number. But that does not really make sense: it would be more comprehensible if the baryon number had started from zero at the beginning, if there had been equal numbers of quarks and antiquarks at that time. Just this scenario is possible if baryon number is not strictly conserved as in SU(5) or SO(10) theory. The new-fangled forces that act in these theories saw to it that the baryon number zero, as it existed at the start, grew to a huge level by our time; that means today's baryon number is a product of history. As we look toward the Universe in its distant future, the baryon number will again assume very different values.

Chapter 11

Conclusion

Electrodynamics and chromodynamics share an important feature: electrons and quarks are point-like objects; they have no inner structure. That means they are on an equal footing as far as elementarity is concerned. There have been attempts to find inner structure in both of them — to no avail. The limit of any possible structure has been pushed down by experimentation at the LEP accelerator at CERN, at the HERA collider at the DESY Laboratories in Hamburg, and at the TEVATRON accelerator at Fermilab near Chicago, and is now in the vicinity of somewhat less than 10^{-16} cm for both quarks and electrons. Still, we cannot exclude some inner structure in quarks and/or leptons, that they consist of even smaller units. Theorists have come up with a variety of models for such structure, but we will not discuss them in the absence of any indication that they relate to reality.

The formulation of the Standard Model of Particle Physics managed to give us a valid description of fundamental forces and particles. But we know well that this does not terminate the search for very basic physics research. We know the Standard Model does not address a number of decisive problems, most urgently the question for the origin of particle masses. The most telling of these, as far as the structure of matter is concerned, are the electron mass of 0.511 MeV and the proton mass of 938 MeV. We can say today that we understand the origin of the proton mass in the context of QCD theory. This theory gives us a simple and evocative picture. The proton mass represents nothing more than the kinetic energy of the quarks and the gluons inside the proton; it is the equivalence of mass and energy given by Einstein's equation $m = E/c^2$. This means that the proton mass, but also the masses of atomic nuclei, stand for the field energy of quarks and gluons they are made of. This notion is based on our understanding of proton substructure: its mass is a direct consequence of the proton's finite extent (at the level of about 1/100,000 of that of a hydrogen atom).

The Standard Model of particle physics tells us that the mass of the electron is due to the coupling of the electron field to the Higgs boson field. The relevant coupling strength is proportional to the electron mass, but is not fixed by any theoretical condition. It is truly eligible — clearly an unsatisfactory situation.

That is why, in order to describe the interaction of electrons and photons, a theory was developed that must be recognized as the most successful theory in existence: quantum electrodynamics (or, for short QED). This theory permits us to calculate interactions between these particles with ultimate precision. To be sure: to do so, we have to include an additional particle, the positron — the antiparticle of the electron, which has precisely the same mass.

The theory takes it for granted that the relevant particles — the electrons, the positrons, and the photons — have no inner structure: an electron is simply a point-like mass subject to the electromagnetic interaction. Not that this signifies utter simplicity: quantum theory tells us that the uncertainty relation implies that empty space is not really quite empty. Rather, it is replete with "virtual" electron–positron pairs-pairs that exist only over minuscule extents in space and time. The vacuum is "filled" with processes of pair creation and pair annihilation.

There are no direct macroscopic consequences of these processes, but they have their influence on the space around the electron: the vacuum will be polarized. Close by the electrons, there will be a preponderance of virtual positrons — which implies that the electric charge of the electron will be partially screened by the cloud of virtual positrons. What follows is this: looking at an electron from the outside, we do not see it as a point-like particle, but rather the electron together with the cloud of virtual particles surrounding it. The proper term for this configuration is a "physical electron", in contrast to an electron devoid of its cloud of polarization; that, by contrast, we call a "bare electron" — a purely theoretical object with a charge that must be larger than that of the physical electron.

Quantum theory thus sees to it that a point-like electron is, in fact, not at all point-like. It looks point-like from a fair distance, but when we look at distances smaller than about a hundredth of the extent of an atom, the effects of vacuum polarization became noticeable, and they are remarkable in the framework of QED. In fact, when we calculate how much of a charge the "bare electron" has in comparison with the measured charged of the "physical electron", we find an outrageous result: it is infinite.

This, in fact, is not the only disagreeable surprise QED has in store for us in this context: similar surprises hit us when we look at the mass of the electron. According to the equivalence of energy and mass, the electric field of the electron must add to its mass — because an electric field means that there is some energy density in the relevant space. Again, the calculation of what amount of mass the field adds, leads to a result that makes no sense: the result is infinitely large. Not that this is hard to understand: in the framework of our theory, we assume that the electron has no inner structure whatsoever, but that it is point-like. That means that the electric field at very small distance is very strong — and a quantized consideration tells us that the field adds infinitely much to the mass. Again: conjecturing infinitely small size as far as the inner structure is concerned leads us to senseless infinities. Not that this puts us to our wit's end: maybe the electron has an inner structure that becomes noticeable only at very small distances — say, at 10^{-18} cm. That would imply the electron is not infinitely small size — rather, it has a very small but still finite radius. It is easy to convince ourselves in this context that this will avoid infinities. Rather, this procedure points out a finite radius of the electron. Notwithstanding a great deal of effort, there has been no experimental hint at some substructure of the electron. Rather, a limit can be set: the radius of the electron, if it is finite, must be smaller than one-hundredth of the extent of a physical nucleon.

In the beginning of the twentieth century, the energy spectrum of the hydrogen atom was found to have a simple structure but remained beyond theoretical interpretation, until the development of quantum theory lifted the veil of secrecy. Now, the mass spectrum of leptons and quarks is in dire need of theoretical penetration. New insight has been gained in particle physics mostly when experimental data indicated a simple structure of the relevant phenomena — and slowly we are reaching this stage of development now.

In the theoretical framework of the Standard Model, leptons and quarks are point-like singularities in space, interacting with fields of force. Does it make sense that those infinitely small "points" are massive? And if so, why is the mass of the muon 207 times that of the electron? After all, these two particles can be distinguished by their masses and nothing else? It may well be that, in the future, we will have to give up the notion of a point-like mass. It might, after all, be that the masses of leptons and quarks are a consequence of a hitherto unobserved substructure of these particles, just as the proton mass is due to its substructure. It cannot be ruled out at this time that substructure effects and hints at new building blocks inside the leptons and quarks will be gleaned from powerful new accelerator experiments.

We have another important reason to take a step back from the interpretation of leptons and quarks as point-like mass singularities. To this day, nobody has successfully integrated gravitation into the Standard Model so as to come up with one unified theory for all interactions. The reason must be seen mainly in Albert Einstein's notion of gravitation as a consequence of the curvature of space and time. While the transition from classical to quantum theory presents us problems to the electromagnetic interaction, the analogous transition is a hitherto unsolved problem for gravitation. Note that a quantum theory of gravitation would have to imply a quantum theory of space and time. That, however, would affect the very foundation on which the Standard Model is built; and changing the concepts of space and time constitutes a real roadblock. Still, we can estimate that at very small distances — say, at about 10^{-33} cm — the uncertainties of quantum theory imply a destruction of our customary notions of space and time. So, if an electron really constitutes a mass singularity, this very singularity will be softened, or smeared, at this minuscule scale. To wit: the implication is that nobody knows what an electron looks like at this scale.

Some theorists conjecture that leptons and quarks are, in fact, manifestations of exceedingly small one-dimensional objects they call "Superstrings". A tiny thread-shaped object is, indeed, less singular than a simple point; and it turned out that the notion of superstrings constitutes less of an obstacle for the formulation of a consistent theory of quantum gravitation. By taking this route, the string theorists have to imply the precondition that our notions of space-time in the Universe — four-dimensional as we know them, with three space and one time dimension — have to be extended to ten dimensions. While this does, at first, look like a serious conflict with our customary observations of space and time, we can arrange the formalism such that only four of those ten dimensions are macroscopically relevant. The remaining six dimensions become relevant only at extremely small distances where gravitational effects come into the quantum game. Let us look at a simple model to illustrate this: we roll a sheet of paper tightly around our axis. From a distance it will look like a one-dimensional object; the second dimension becomes noticeable only when we look at distances commensurate with the radius of this roll of material. And analogously we can imagine that six ulterior dimensions have been "rolled" such as to become unnoticeable.

Should it be true that, at very small distances, our three-dimensional space actually acquires more dimensions, we might even use these to work on problems in particle physics. For instance: to this day, we have no idea why Nature prefers the number 3. There are three families of leptons and of quarks, and the quarks in turn have access to total of three "colors". It might be that these symmetry structures relate directly to the hidden dimensions, and that by inference the phenomena of elementary particles will find geometric explanations. Active research is following up on such possibilities.

To sum up: the many successes of elementary particle physics demonstrate to what astonishing degree we are already understanding our world. It appears that when dealing with the smallest constituents of matter, we will finally hit on mathematical formulae, as the Greek philosopher Plato predicted millennia ago. Science, philosophy, and ethics wind up at a common meeting point.

Whether such a geometric approach to the physics of elementary particles will ultimately be successful remains, at out time, a secret, just like the structure of the mass parameters of leptons and quarks, tiny, tiny particles in the Standard Model, which in reality may not be so tiny after all. This unsolved set of ultimate questions is already casting its shadows into our new millennium — reminding some of us of Bertold Brecht and the note he jotted into his diary in 1921: "If there are no mysteries, there cannot be any truth".

For promotional purposes:

This book describes the physics of elementary particles for a broad audience. As an introduction, the physics of atoms and molecules is discussed, followed by a description of the quantum nature of atoms and particles. Then the quarks are introduced as the elementary constituents of atomic nuclei. The basic ideas of quantum electrodynamics and of quantum chromodynamics, as well as the theory of strong interactions, are then discussed.

The nuclear particles are interpreted as bound states of quarks. We then focus on the gauge theories of the weak and electromagnetic interactions, as well as on the problem of mass generation. At the end of the book, the ideas of grand unification of all basic forces are described, based in particular on large symmetries like SU(5) or SO(10). Finally, we discuss briefly some astrophysics applications. This book is suited for a reader who is interested in physics and knows a little bit about science — say, on the high school level.

The author played an important role in developing the theory of quantum chromodynamics, together with Murray Gell-Mann, who introduced the idea of quarks, and also contributed to the concept of grand unification.

Index